T0074082

Cognitive Systems Monographs

Volume 17

Editors: Rüdiger Dillmann · Yoshihiko Nakamura · Stefan Schaal · David Vernon

Lidia Ogiela and Marek R. Ogiela

Advances in Cognitive Information Systems

 Springer

Rüdiger Dillmann, University of Karlsruhe, Faculty of Informatics, Institute of Anthropomatics, Humanoids and Intelligence Systems Laboratories, Kaiserstr. 12, 76131 Karlsruhe, Germany

Yoshihiko Nakamura, Tokyo University Fac. Engineering, Dept. Mechano-Informatics, 7-3-1 Hongo, Bukyo-ku Tokyo, 113-8656, Japan

Stefan Schaal, University of Southern California, Department Computer Science, Computational Learning & Motor Control Lab., Los Angeles, CA 90089-2905, USA

David Vernon, Khalifa University Department of Computer Engineering, PO Box 573, Sharjah, United Arab Emirates

Authors

Lidia Ogiela

AGH University of Science and Technology
30 Mickiewicza Ave, 30-059, Kraków, Poland
E-mail: logiela@agh.edu.pl

Marek R. Ogiela

AGH University of Science and Technology
30 Mickiewicza Ave, 30-059, Kraków, Poland
E-mail: mogiela@agh.edu.pl

ISBN 978-3-642-25245-7   e-ISBN 978-3-642-25246-4

DOI 10.1007/978-3-642-25246-4

Cognitive Systems Monographs   ISSN 1867-4925

Library of Congress Control Number: 2011942876

Typeset by Scientific Publishing Services Pvt. Ltd., Chennai, India.

Printed in acid-free paper

5 4 3 2 1 0

springer.com

Preface

The progress in computer science is now mainly achieved by scientific circles and constitutes a kind of indicator of the position of centres dealing with this area. This is because nowadays one cannot talk of highly developed scientific units, economic growth or world-class achievements in various disciplines of knowledge if one does not conduct research in computer science (whether technical or mathematical). The development of computer science is now so rapid that we, the readers, increasingly receive technology news about new solutions and applications which very often straddle the border between the real and the virtual worlds. Computer science is also the area in which cognitive science is witnessing a renaissance, because its combination with technical sciences has given birth to a broad scientific discipline called cognitive informatics. And it is this discipline which has become the main theme of this monograph, which is also to serve as a kind of guide to cognitive informatics problems.

This book is the result of work on systems for the cognitive analysis and interpretation of various data. The purpose of such an analytical approach is to show that for an in-depth analysis of data, the layers of semantics contained in these sets must be taken into account.

This approach to this subject was made possible by work to combine the subjects of intelligent information systems and the cognitive aspects of the human analysis process. The interdisciplinary nature of the solutions proposed means that the subject of cognitive systems forming part of cognitive informatics becomes a new challenge for the research and application work carried out.

The authors of this monograph hope that it will guide Readers on an interesting and accurate journey through the intricacies of information and cognitive science. Thus it may make us, when we look at the world around us, wonder (sometimes jocularly) whether we have really got to know it, whether we understand it, and whether we will ever be able to accurately and unanimously (avoiding contradictions) explain what happens around us.

Lidia Ogiela
Marek R. Ogiela

Contents

1 Beginnings of Cognitive Science 1
   1.1 Cognitive Problem Development...10
   1.2 Cognitive Methods Dedicated to Research and Practice 17

2 Fundamentals of Cognitive Informatics 19
   2.1 Formal Cognitive Models 20
   2.2 Cognitive Resonance Model...32
   2.3 Semantic Analysis 35
   2.4 Semantic Categorisation 41
      2.4.1 Using Semantic Categorisation to Analyse Speech and
             the Speaker...43
      2.4.2 Semantic Analysis in Medical Systems for Cognitive Data
             Interpretation...48

3 Cognitive Information Systems 51
   3.1 General Classification of Cognitive Information Systems.....................52
   3.2 A Formal Perspective on Cognitive Categorisation Systems.................57
   3.3 Properties of Cognitive Categorisation Systems....................................59

4 Intelligent Cognitive Data Analysis Systems of the UBMSS Type
   as an Example of Cognitive Categorisation Systems 61
   4.1 A UBMSS System for a Single-Factor Analysis of the NPV 63
   4.2 An Example UBMSS System for a Single-Factor Analysis
        of the IRR 66
   4.3 An Example UBMSS System for a Dual Factor Analysis
        of the IRR and the Discount Rate r...69
   4.4 An Example UBMSS System for a Multi-factor Analysis
        of the Economic and Financial Ratios 71

5 UBIAS – Intelligent Cognitive Systems for Visual Data Analysis 75

6 E-UBIAS – Cognitive Systems for Image and Biometric Data
   Analysis...85

7   Cognitive Systems and Artificial Brains 99

8   Summary 107

References 109

Internet Sources 117

Index 119

Chapter 1
Beginnings of Cognitive Science

The first mention of cognitive science can be found in the works of Aristotle, who proposed two dominant categorisation methods describing all varieties of cognitive science in different ways. Aristotle's considerations, also of the concept of a category, led to distinguishing accidental and substantive categories based on the differences Aristotle saw between the subject of a sentence treated as the substance and the predicate treated as an accidental category. The substantive category includes concepts that describe something and present something concrete, so they were a 'concrete substance', the subject of a sentence, something material. Within the accidental categories, Aristotle distinguished nine basic notions, which included quantity, quality, relation, place, time, location, property, action and sensation.

Aristotle's considerations gave birth to a method currently referred to as 'top-down', which defines a concept based on the type (genus) and the appearance of a single or several differences (differentiae) allowing new genera of forms from to be distinguished other forms of the same genus. This type of propositions Aristotle formulated in his works on logic, but in those on biology he criticised the limitations of the top-down approach and at the same time proposed an approach currently called 'bottom-up', which starts with detailed descriptions and definitions of an individual, classifying collections of individuals into species and genera and grouping different genera in groups. Aristotle considered the top-down method to be right for presenting and describing the results of his analyses, reasoning and proofs on this method, but he clearly favoured setting apart the bottom-up method as better for discovering subsequent research procedures on a new object.

In the third century after Christ, Aristotle's considerations of a concept, definition and category became the foundation of the work by Porphyry[1], who undertook the effort of writing a commentary to Aristotle's categorisation. This commentary contained the first notes on a tree diagram shown in Figure 1.1.

This tree shows categories and references to syllogisms with Aristotelian laws and rules concerning reasons associated with genera and types of defined subgenera.

The diagram shows an ideal genus, genera, lower levels, subgenera, the closest level, a species and an individual.

---

[1] Porphyry, a recognised ancient Napoli philosopher and astrologist. Known as commenter of Plato's and Aristotle's works, edited the works of the ancient philosopher Plotinus. Porphyry became famous for his collection of commentaries written to Aristotle's *Categories*, known as *Isagoga*, which constituted the most important text on logic in the Middle Ages and formed the starting point of research on logic and the dispute about universals. Porphyry's commentaries showing the dependencies of species and genera presented using a tree became the cornerstone of contemporary taxonomy.

L. Ogiela and M.R. Ogiela: Advances in Cognitive Information Systems, COSMOS 17, pp. 1–18.
springerlink.com   © Springer-Verlag Berlin Heidelberg 2012

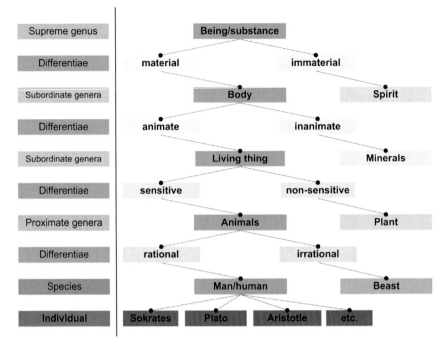

Fig. 1.1. Porphyry's tree presenting the diagram of Peter of Spain (1239/1947). Source: developed on the basis of [24]

In this diagram, one can observe how an ideal genus called the 'being' moves to a lower level, as a sub-type and one of genera of this being turns into a being called the 'body', and at the same time a non-material genus of a 'being' at a lower level turns into the sphere of the 'spirit'. The entire technique of heritage presented in the process shown in the Porphyry's diagram also combines other categories, which include living organisms, defined as animate 'beings' made of matter and a 'man' defined as a 'being' that is rational, sentient, living and made of matter. Porphyry uniquely points to representatives of the human species, in particular the known fathers of philosophy: Socrates, Aristotle and Plato.

Cognitive science developed further in the times of Blessed Ramon Llull, the first to propose a concept of the machine application of sentences for simple categorisation jobs. It is worth noting that he did this as early as in the 13th century.

Ramon Llull invented a mechanical device, a logical machine, in which the subjects and predicates of theological statements were arranged in circles, squares, triangles and other geometrical figures and when a lever was thrown, a crank handle or a wheel turned, these statements formed true or false sentences, proving themselves. Llull called this device *Ars Magna* [66] and devoted his most important works to its description. His work, *Ars Magna,* was physically a set of inscribed discs with primary concepts which could be combined in various ways by the appropriate rotation of these discs (Fig. 1.2).

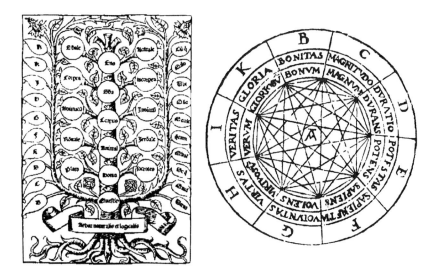

Fig. 1.2. *Ars Magna* by Ramon Llull. Source: [192]

His plan was founded in theoretical philosophy, or rather theosophy, as the primary motive for Ramon's method was that theology was identical with philosophy. He claimed that there was no difference between philosophy and theology, reason and faith, so that the highest secrets could be proven using logic and his machine – the *Ars Magna*. The best known edition of the work in which Ramon Llull described his logical machine was the Strasbourg edition of 1651.

Ars Magna has also been the subject of contemporary work, and one of the most often cited versions is an attempt at the automatic implementation of Llull's solutions made by Steven Abbott and Yanis Dambergs of 2003 (Fig. 1.3).

Llull's system originally inspired Gottfried Wilhelm Leibniz to propose his idea of *Characteristica Universalis* [64], presenting very simple sentences using numbers and at the same time building sentences resulting from its operation.

Expressions like "All *A*s are *B*" were checked by holding the number assigned to the concept *A* and checking whether that number was divisible by the number assigned to the concept represented by the letter *B*. If the noun 'flower' was represented by number 12, and the adjective 'creeping' by the number 23, their smallest common multiple was treated as a common conceptual category. In the example problem, the result – the number 276 – represented a 'creeping flower'. If 'grapevine' was represented by the number 11,316, then the sentence 'All grapevines are creeping' was true, as the number 11,316 is divisible by 276. Today, looking back, Leibniz can be said to have proposed a universal dictionary capable of converting words, sentences or syllogisms into numbers which could be the subject of reasoning based on the rules of arithmetic. To simplify the calculations necessary for his method, Leibniz also proposed the first computing machine capable of multiplying and dividing.

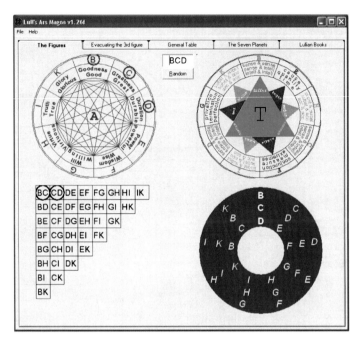

Fig. 1.3. The implementation of Ramon Llull's *Ars Magna* by S. Abbott and Y. Dambergs. Source: [185]

Thus computational linguistics begun with the introduction and implementation of Leibniz's universal dictionary. It is noteworthy that this method is still of interest today. One example comes from 1961, when for the purposes of a machine translation system, Margaret Masterman, a former student of Ludwig Wittgensteins, defined a network of 15,000 words based on compounding 1,000 simple, basic concepts [70]. In 1975, Roger Schank reduced the number of primary actions that had to be executed when using computational linguistics to just 11 [125]. According to his concepts, the transformation of high-level concepts into simple, basic ones, had to use two different phrases synonymous to one another.

When considering methods of computational linguistics, it is worth remembering that the system generally used in it must allow the transformation of high-level (complex) concepts into concepts (words) of a lower level. Such a system must also have a constructive component enabling the construction of complex concepts from correctly combined simple concepts, but this enhancement of the system, treated as a certain change of the basic rules, must be optional, not obligatory.

All the methods mentioned above were based on logical foundations. They used a logical apparatus for their operation, starting with Porphyry's tree down to formal ontologies, exemplified by Aristotelian top-down approaches. However, let us remember that Aristotle himself preferred the use of a bottom-up method for empirical data analysis. In 1858, William Whewell extended the definition of the top-down method, making it more precise and dedicating it to natural science [149]. In 1865, John Stuart Mill proposed conjectures and defined the final conditions for creating a 'closed form', while still believing it possible to define a 'closed form' [74]. He was an advocate of proposing criteria based on necessary characteristics and suggested the optional selection of those of them which, after a given theory was proven, could serve effectively. Various phases can be distinguished in J. S. Mill's works, in which he professed various beliefs and preferred various methodologies. During the period in which he supported utilitarianism, he detailed and adjusted the concept of the empirical theory of cognition, several thoughts and theories from which have become the foundations of today's mathematical logic. The most important of them undoubtedly include the first attempt to narrow down the concepts of traditional Aristotelian logic. Mill presented the rules, which in Aristotle's works had the form of unclear narrations, as diagrams, schematics and symbolic representations. In addition, he proposed to stop elevating logic and mathematics to a superior level and instead view both of these sciences as languages which can be freely, but in a non-contradictory way, adjusted to the needs of their users. Important achievements of Mill's also include the attempt to structure inductive reasoning rules into so-called Mill's principles: methods, whose use led to producing categorical judgements by way of induction.

Mill's methods are as follows: the method of agreement, the method of difference, the method of residue, the method of concomitant variations and the joint method of agreement and difference. These subjects are too broad and too detailed to discuss them in this book, but should be suggested to the Reader as the direction of further study. The progress in Mill's thinking and views meant that this philosopher later moved closer to naturalistic views of the theory of cognition, while in ethics he went in the direction of neo-Kantian theories. However, this subject goes far beyond the limits of this book.

For centuries, philosophy and science struggled, with greater or smaller success, with attempts that were to produce a prototype learning system and at the same time to create a general cognitive system that could be used to describe the human process of acquiring knowledge, but without losing the ability to describe similar processes taking place in an artificial, computer system. Every great scientist tried to relate to the theories and theses of his equally great predecessors as well as extend and improve those theories, whose aim was to get better knowledge of not just the human being, but also the entire world around it.

Over the last centuries, methods aimed at the formalised writing of definitions, logical methods, fuzzy methods and prototype building have developed greatly. All the above methods are aimed at defining words, concepts, meanings etc. Even though they had already been proposed in the 19th century, they were then effectively implemented also in modern computer systems.

The progress of civilisation, particularly the western one, has for many years been accompanied by the belief that only correctly conducted scientific research could ensure progress in knowledge. This conviction also applied to the processes of cognition (and understanding) treated as the subject of research at the time when the toolbox already developed for science was applied to studies of these phenomena. The pioneers of this approach were Hermann von Helmholtz [44],

Wilhelm Wundt and Gustav Theodor Fechner, who first initiated psychology understood as a separate science [92].

Of the many fields of psychology in existence today, this book will refer mainly to cognitive science, an experimental discipline of knowledge studying various manifestations of the cognitive activity of the mind. When cognition and understanding moved from the domain of poetic metaphors to scientific exploration, it halfway became a typical research subject, though difficult to theoretically conceptualise and empirically test. It soon became clear that the scientific research of cognitive science problems required the involvement of many disciplines, so researchers tried extending the scientific approach to studying the mind. This was helped by the appearance of computational disciplines, including cybernetics (Norbert Wiener) [150], [151], information science (John von Neumann) [140], [141], artificial intelligence (John McCarthy, Marvin Minsky, Lotfi A. Zadeh) [71], [76]-[78], [154]-[157], as well as broad-ranging research of the brain (Donald Olding Hebb, David Hubel, Torsten Nils Wiesel) [42], [43], [47], [48] and theories of generative grammars in linguistics or of formal grammars (Noam Chomsky) [19], [20].

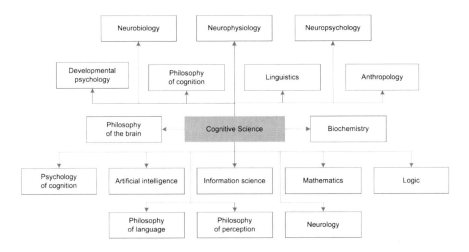

Fig. 1.4. Scientific disciplines making up cognitive science

Cognitive science straddled the borders of various scientific disciplines (Fig. 1.4) which formed its foundations, which inspired it, and on whose basis it started developing and gaining in importance.

It was quickly noted that the results obtained in individual disciplines were important not just for the narrow group of specialised researchers, but that a new, unusually original subject was arising, which could be defined as a systemic perspective on the mind, which treated cognition and understanding as a multi-stage computational process. In the cognitive science view, every mental process is treated as information processing. This processing is then characterised as parallel and distributed, as this is exactly how information is handled in systems having the architecture of neural networks. Obviously, the computational representation of cognitive processes really boils down to developing computer models of these complex biological and psychological processes. However, due to the complexity of the object studied, this modelling approach seems particularly convenient, as models can not only be created, but also researched experimentally using computer simulations. The adequacy of these models is confronted with the neurobiological knowledge of the structures and functions of nervous system parts and elements, constantly collected and extended thanks to new research techniques of brain biology. This approach has meant that a 'cognitive science revolution' begun in the development of science, resulting in the creation of a pure science of cognition – cognitive science.

One of the reasons for the development of cognitive science was the recognition of the fact that the problem of cognition turned out to be much more complex than originally thought, and getting to grips with its intricacies implied a new split of research work between those collecting empirical facts and those who interpreted them. What started distinguishing cognitive science from other sciences (philosophy or psychology), was its multidisciplinary nature. The term 'multidisciplinary' was introduced to characterise fields of science (such as cognitive science) which are founded in many other fields and which significantly contribute to the development of these fields. This term was introduced to mark a difference from 'interdisciplinary', which appears when some well-defined research or engineering problem requires the cooperation of representatives of various scientific disciplines, each of which contributes the tools typical for his/her discipline, but does not receive much in return[2]. The situation was different in researching cognition and understanding processes, where detailed disciplines contributing to the above multi-disciplinary approach receive contributions to their own development and growth from the common resource to a greater extent than they contribute ready solutions to that resource.

Looking back, it has been observed that today's science can distinguish and research only small fragments of the complex set of problems associated with cognitive science. Such isolated cognitive science problems as perception, imagination, memory, learning, conceptual thinking, understanding and many others have been researched – but each one separately. Every one of the above forms of cognitive activity of the mind is now the field of interest of a separate research discipline, and in addition, each one can be analysed at many levels. The most widely used structure of these levels has been described by David Marr [69], who distinguished three primary levels:

- the computational (theoretical) level,
- the level of representation and the algorithm (the interface between the theory and the empirical),
- the hardware (mainly computer) implementation level.

---

[2] Such work was done to, for example, construct the nuclear bomb, space rockets or analyse the human genome [91], [92].

Cognitive science is today pursued by, among others, neurobiologists, psychologists, linguists, sociologists, computer scientists and philosophers, and every one of these scientific circles obtains something different, valuable and useful from it.

The efforts that contributed to the development of modern cognitive science were mainly empirical projects. Inspired by their intention to study the nature of the human mind, people have experimented for years on animal brains and conducted other research based on empirical data about pathologies of human brains as well as psychological observations. This led to distinguishing a specific neuroscience with many important achievements and discoveries to date. In another area of cognitive science, sentences assessed as correctly built were subjected to linguistic analyses on the assumption that a significant part of human cognition is contained in linguistic communication between people (every learning must be communicated inter-subjectively), and in internal verbalisation which forms the foundation of thought processes conducted at a certain level of abstraction. The nature of cognitive processes was additionally clarified by artificial intelligence, with its designs of machines or their associated programs that could undertake cognitive activities.

Thus cognitive science started to look like a synthesis of knowledge about the mind, including philosophical reflections on its nature, knowledge about psychological phenomena and rules governing the behaviour of people and animals, complemented with the study of the language, the biological foundation for psychological phenomena, and also making use of their cybernetic interpretation. Thus cognitive science encompasses research on the brain and on psychological manifestations of its operation, but also mathematical models and an engineering approach making it possible to build a technical structure similar to a human brain and at the same time to try to use it for teaching and improving this brain. The basic areas which cognitive science deals with are most often considered to cover [92]:

- the impact of mega-information on human behaviour;
- the impact of micro-information on human behaviour;
- the emotional modelling of human behaviour;
- the cultural modelling of human behaviour;
- matter vs. the spiritual world;
- human impressions and their quality;
- functions of the consciousness;
- functions of the brain;
- formal systems and the meaning of symbols;
- neural models;
- technical aspects of artificial intelligence;
- mind vs. brain.

Due to the above features, cognitive science covers very diverse areas of work, including the following [92]:

- attempting to explain the mechanisms of perception;
- man-computer interactions;
- optical illusions;
- acoustic illusions;
- processes of learning supported with multi-media technologies;
- aspects of human memory;
- development of memory processes;
- teaching processes;
- knowledge acquisition;
- information analysis and interpretation;
- knowledge structure.

Cognitive science is applied very broadly, because it concerns the development of computational intelligence, including methods of machine recognition and classification of patterns, multi-dimensional statistics including subjects such as clustering and discriminating, probability-based methods etc. Cognitive science can also include research on optimisation and the modelling of uncertainties, research in areas close to technical cybernetics such as sets and approximate logic, as well as the theory of quality control and modern control technologies.

Cognitive science problems are increasingly frequently dominating not only the development of humanities, but also science, because cognition models are becoming the pattern, the purpose and the basis for the operation of various types of systems, including information systems designed for:

- business;
- economy;
- logistics;
- automation;
- cryptography;
- medicine;
- aviation;
- government institutions;
- transport;
- commerce;
- industry.

They are developing incessantly and extremely rapidly (which is necessitated by the growing economic needs) in many directions, but to an increasing degree this is founded on cognitive analysis. This is why most specialists deploying computers at companies are becoming convinced that cognitive analysis methods will soon become extremely useful for understanding economic situations, which will improve the operation and contribute to the development of computerised management systems.

1.1 Cognitive Problem Development

Cognitive subjects are of interest to various scientific disciplines stemming from different theoretical and methodological traditions. Cognitive science is very often understood as the science of learning and is therefore identified with the cognitive theory. Cognitive theories started developing with the introduction of the philosophical cognitive theory - epistemology [158]. Representatives of epistemology focused their work on fundamental problems of human learning, which included the sources and the nature of knowledge acquired as well as the theory of the truth. Thus epistemology dealt with the human cognition, and in particular its:

- subject – everything that exists, every object or being that can be known in a way concerning an individual person, but also 'discovered'; whose properties, individual features, meaning etc. can be seen;
- content – the meaning of the object or being learned about (its semantics);
- processes – related to learning, the definition of a given concept, the definition of signs that can be assigned to given objects, creating analogies, selecting the language, formulating judgments on the objects (beings) learned about etc.
- methods;
- limits;
- criteria.

Cognitive theory was at the beginning developed as a speculative science supported by logic and everyday observations. However, as time passed, it increasingly started to use the achievements of philosophy, psychology, medicine, linguistics and also informatics. Cognitive science, as the science of learning, started focusing on a special field of science - the philosophy of the brain, which then studied the classical psycho-physical problem originally defined as the relationship between the tangible body and the intangible soul, later understood as the relationship between the brain and conscience. At the time, cognitive theory concentrated on matters explaining how the brain can generate conscious conditions of the psyche and on determining the function and the origin of qualia - qualitatively individual, subjective states of consciousness.

In addition, in-depth observations were made of the expression of emotions in people and the phenomenon of emotions arising from the perspective of the human evolution. Foundations were laid for the development of a scientific discipline based on researching the phenomenon of emotions in a fashion that would open the way to studying it experimentally, and scientists wrote about the significance and the role of emotions for psychopathology.

The issue of emotions and the entire phenomenon connected with them had a huge impact on the appearance of two existence models of mental processes - the neuro-physiological and the perception model. Each one of them is present in the human brain to a greater or smaller extent and is based on emotional states triggered earlier, starting at the initial phases of the cognitive analysis process. For many years, at many scientific centres world-wide, research was carried out on the phenomenon of emotional states occurring and their link to basic functions of human mental processes making use of attention, perception, memory or language [24]. The 20th century saw critical reflection about the development of cognitive science, namely that work aimed at studying the impact of emotions had been totally neglected. As a result, this somewhat neglected area became the subject of intense research. It is notable how small the group of scientists was who researched emotions from the perspective of cognitive science, but who laid the foundations for an extremely important analysis of this phenomenon and at the same time characterised many components and tremendously important details of the researched phenomenon. They were: Paul Ekman [33], Jerome Kagan [51], Richard S. Lazarus [61], George Mandler [67], Robert B. Zajonc [158], Steven Schachter and Joel Singer [124].

Emotional processes studied by the above scientists formed firm ground for learning the functions and properties of a psychogenetic area called the limbic system. The links between that system and the emotions that arise are of great importance because emotions are compared to other functions connected with other brain structures. The difference between emotions and the cognitive acquisition of new information is due to the duality of structures and functions between the limbic system and the neocortex.

A new element in the research work on human emotions was the presentation, by many researchers studying neurophysiologic cases, of extremely important links between human emotional states and the right brain hemisphere (Roger Sperry, Joan C. Borod [11], Richard Davidson [25], Steven Z. Rapcsak, Joel F. Comer and Arnold E. Rubens [121]).

Psychology distinguished the following stages in the process of understanding any information received by a human and subjected to a cognitive analysis:

- *information recording* – may boil down to a single perception cycle, but it may also take a complex form which assumes an explorative activity of the individual;
- *memorising* – may consist in the simple fixing in memory of detailed information, i.e. facts, patterns of objects and methods of action, but in more complex situations may consist in creating a universal memory trace (a gnostic unit) capable of generating features subsequently needed to use the knowledge possessed in the process of understanding new situations being analysed;
- *coding* the information obtained – making it confidential, its encryption and decryption form the basic components of the coding phase. It is also possible to split information and share it;
- *storage* – the latent stage of mental processes which cannot be researched directly, whose nature and course must be reasoned out from the last phase; There are reasons to believe that storage is not just a passive process of keeping information, but consists in creating subsequent versions of more and more refined internal representations of the knowledge possessed and systems of its internal links;

- *retrieving* information – covers remembering, recognising, understanding and re-learning specific skills. Retrieval is the moment at which the acquired knowledge and wisdom resulting from it may manifest itself in the individual's behaviour.

Useful knowledge, which among other functions facilitates generalising the information learned usually takes the form built at the memorising stage.

The retrieval stage is the measure of the human memory processes: the remembering, recognising, understanding and possibly learning anew. Another differentiation consists in linking memory as a characteristic of the nervous system to individual analysers. Memory dependent on the analyser type can be called a peripheral ability. For the essence of mental processes it is important that memory also exists as a general ability, working when complex stimuli act, achieved as a result of the operation of many analysers. The most important features of memory used to assess this mental process are:

- permanence – a criterion related to the storage stage;
- speed – defined as the ease of recording new facts and links between them; this criterion belongs to the memorization stage;
- accuracy or reliability – defines the relations between the retrieved information and what constituted the contents of the memorization stage;
- readiness – adequate to the retrieval stage – determines whether remembering occurs without major problems or whether additional activating stimuli are necessary;
- range or capacity – refers to the coding phase.

Recognition, which is one of the indications that memory works, is inseparably linked to perception and decision-making. When perceiving objects in their surroundings, individual either recognize them as known (they can name them and associate them with a specific action), or find that they do not recognize them, which also plays an important role as the so-called detector of novelty.

Both if the specific object is recognised and if it is detected as a novelty, the human brain assigns a specific mental category to this perception situation, and thus takes the decision whether the stimulus acting upon it is old or new, known to it or unknown.

This mechanism is of fundamental importance in the entire psychology of perception. The sensory threshold – of susceptibility and sensitivity – is determined by reference to the recognition process. A threshold stimulus is received or recognized as a stimulus in 50% of cases, and in 50% of cases it is not. Similarly, the difference threshold is set at that value of the stimulus which in 50% of cases is recognized as equal in value to the standard stimulus, and in 50% as different.

The neurophysiologic model of cognitive analysis is based on the operation and behaviour of the brain which can be described by studying attractors resulting from the dynamics of great groups (in the order of millions) of neurons. These attractors are defined in the stimulation space of such neurons. Even though the surface states of the brain (e.g. their activity examined using the EEG) are not identical, in the dynamics of the brain function we can find constant relations between "practical attractors" which are relatively stable set neurophysiologic states that can be identified in brain behaviour. What is more, dynamic brain states, if they are properly handled mathematically, are characterised by certain deeper relations which may have a simple logical representation.

The correspondence between the mental states and forms of brain activity does not concern only surface states which are of an ephemeral (transitory and short-lasting) character, but also attractor states which exhibit some stability. However, internal representations, which according to model research must be more stable, do not translate into similarly stable surface representations, which makes their empirical study difficult.

In addition, small disturbances in the electro-chemical structure of the brain cause significant changes at the mental phenomenon level, e.g. bring about mental illnesses. These are frequently the consequence of the global brain dynamics, dependent on the condition of all its structures, and in most cases no single, well defined anatomical structure responsible for the specific form of psychopathology can be pinpointed. In the brain, a holistic principle can generally be observed which consists in all structures contributing to a greater or smaller extent to all functions. This leads, *inter alia,* to a fact, known for many years, that the brain has no areas in the anatomic sense which are interconnected so weakly that they could be eliminated (e.g. excised) without changing its function, although sometimes those changes may be difficult to detect.

This justifies the claim by John R. Searl [126], that although the neurophysiology of the brain does determine the character of the mind, this does not mean that intelligence or some form of a mind could not be created by the interaction of elements based on silicone or other, non-carbon compounds. The development of computer science, and in particular artificial intelligence, has proven that intelligent information processing is also possible using silicone, and the neurochemical states determining human mental impressions can, in turn, be approximated with a satisfactory precision using digital devices. However, such 'minds' have a structure different from that of human minds, which means that the 'way in which they experience the world' is different because different brain structures must bring about different minds. So the basic features that a given system should exhibit so that its aspects can be equated to that of a mind include intentionality, understanding and conscience. At the first glance, these features seem to describe strictly biological systems, but the evolution in the behaviour of robots and IT systems has made them of interest to technology as well. We now have many examples of systems controlled by software that features 'brain-like' methods of information processing. Another feature of these systems is that only their general properties are programmable, which allows them to be considered as moving towards intentionality. We currently observe the appearance of a number of technical solutions featuring an increasing autonomy of operation and it seems that the convergence limit of this series of better and better models may be intentional systems with increasingly complex structures and purposes of operation.

The progress in neurophysiologic knowledge increasingly convinces us that it is the structures of brains that are the reason and the drivers of the ways in which minds operate. Those structures demonstrate the obvious dependency of the complexity of mind forms on the complexity of the brain itself. What is more, the increasing complexity of the mind and the brain is evidenced by complex forms of behaviour. It seems indisputable that the increasingly professional computer models created using neurophysiologic data contribute to a more and more detailed explanation of aspects of the mind's operation. Every self-organising system currently exhibits intentional behaviour coming from the pursuit of general values and needs of this system which result from its biological structure or technical design. In the process of developing such models of information systems which allow information to be processed and understood in a way modelled on the human brain function, we should now expect increasingly complex forms of behaviour.

The human brain is the location of real neurophysiologic processes which cause real, conscious impressions. Many of the above processes are executed by areas of the brain that have now been well located and are observable using various techniques, combining morphological imaging with functional monitoring, for instance the positron emission tomography [92]. This allows us to study processes during which several sensory modalities must cooperate and references must be made to episodic memory layers located in a specific area of the cerebral cortex. The most frequent ones are analyses of sensory data, such as segmenting an image. Components kept in the episodic memory contain the aspects of a specific modality that cannot be executed locally. What becomes necessary for the whole process is a mechanism for distributing information and combining results obtained from various areas of the brain into one whole. This integrating operation of the brain means that perception processes use fragments of mental representations supplied by the senses and internal stimulations of areas processing sensory information, or a holistic representation of a complex situation containing all the fragments currently stimulated. This is a perceptual meta-representation of this complex situation, probably produced by the global dynamics of bioelectric discharges in the brain, which can be observed as β waves on the EEG.

Cognitive science concepts are useful as far as they approximate models describing the brain's operation at the neurophysiologic level. If some regularities can be found in neurophysiologic observations, an attempt is made to express them as certain rules of logical procedure which can then be mapped to cognitive science models. Consequently, in some models of information systems, it is convenient to use the notions of classical cognitive science, primarily logical rules used to explain mind states in order to simplify their biophysical models. This paradigm can be related to many levels of the description of neural structures and functions, which include:

- the sub-molecular, or the genetic, level;
- the molecular cell structure level;
- the level of neurochemical phenomena;
- the level of single neuron activity;

- the level of activity of neural nuclei and other, anatomically distinguishable areas of the brain;
- the level of the global dynamics of bioelectric phenomena in the brain which correlates with states of mind.

It cannot be disputed that for a long time yet the simulated mind will not be identical to the biological one, *inter alia* because of the technical difficulties in this type of simulations. Currently, the accuracy limits in simulating models of information systems on computers are not completely clear. The majority of models broadly used in neuroinformatics are very simplified, like the popular, widely used neural networks. However, attempts to model neural systems realistically (e.g. de Schutter's *Genesis* project) prove that computer techniques can also map neural systems with a precision and fidelity which had until recently seemed completely impossible.

The naturalistic solutions presented above lead to various empirical predictions and in the future may form the foundations for developing a satisfactory theory of the brain.

The development of cognitive subjects concerned not only social sciences like philosophy or psychology: they developed no less successfully in the fields of informatics and mathematics, where definitions of mathematical concepts, notation and symbols played a significant role.

Artificial intelligence, which leads to designing computer systems capable of executing individual cognitive actions, represents a special approach to the study of cognition. Artificial intelligence started focusing on engineering sciences, where intelligent systems were developed for practical use. Their designers strove to enable these systems to imitate people's cognitive processes very effectively. The success of systems built was assessed by reference to their practical utility and the possibility of deploying them to conduct basic research on cognition. The term 'computer modelling' appeared and was accepted as one of the basic research methods used in cognitive psychology. In the light of the extremely interdisciplinary research conducted, people started talking of a new field of research, not just on cognition, but also on understanding and reasoning - the cognitive science.

In 1978, Alfred P. Sloan Fundation wrote a report in which he stated that cognitive science was an attempt to combine subjects followed by six basic, traditional scientific disciplines [24], [92]:

- philosophy;
- psychology;
- linguistics;
- informatics;
- anthropology: philosophical, biological and sociological;
- neural science.

All the above disciplines cooperate within cognitive science in line with the scientific links between them, presented in Fig. 1.5.

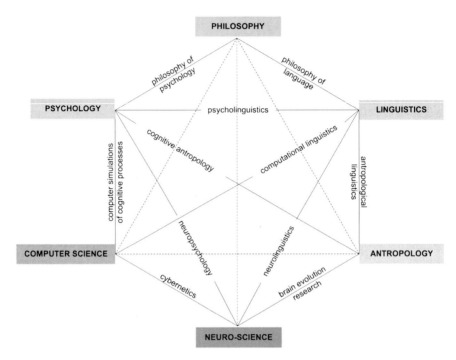

Fig. 1.5. Sciences making up cognitive science and links between them. Source: developed on the basis of [6], [24]

The history of cognitive science has seen the gradual introduction of new problems represented by novel mathematical forms into this science [8], [119], [155], [157].

The use of mathematical tools in cognitive science boiled down to two categories which could be referred to as analytical and semantic mathematics [87]. Mathematical structures based on analyzing variable functions and operations were used to build system architectures analysing data in dynamic processes.

The analysis of this type of events led to distinguishing a new branch on the border of cognitive informatics, computational intelligence, software engineering, knowledge engineering and the application of mathematical forms of definitions of the analysed problems. In recent years, Wang has called this field denotational mathematics [142]-[144]. The key forms and structures of the mathematics created by Wang are:

- the notion of algebra;
- the system of algebra;
- the algebra of real time processes;
- the semantic algebra.

These four basic types of mathematical structures are used in practice to execute tasks not just in the area of cognitive informatics, but also in computational intelligence or knowledge engineering.

Apart from Wang's work illustrating the use of cognitive methods based on mathematic formalisms in computer science, work is also conducted to use cognitive subjects for the computer analysis of data. This type of application of cognitive structures largely leads to the structural and semantic analysis of data under consideration. Structural analysis is based on applying cognitive methods to the information about the structure, the form and the shape of the analyzed object. On the other hand, semantic analysis uses the information contained within the data: the semantic information (meaning).

1.2 Cognitive Methods Dedicated to Research and Practice

Cognitive methods and solutions are important in computer science work, as computer science has become a scientific discipline in which cognitive science is currently blossoming. It is found at almost every stage of the development and the practical implementation of new solutions, which include:

- fingerprint scanning – biometric analysis;
- speech recognition – voice analysis;
- speaker recognition – personal analysis;
- face recognition – identity analysis;
- DNA analysis – identity analysis;
- image analysis – recognition and reasoning analysis;
- signal analysis – sound analysis;
- ratio analysis – decision-making analysis;
- automatic control – motion analysis;
- decision-making analysis.

The above types of cognitive analyses are used in diverse areas, which include:

- airport control (biometric analysis, face recognition);
- computer system login (face analysis, speaker analysis);
- identification of the deceased (DNA analysis);
- interpretation of medical diagnostic tests (image analysis);
- automotive traffic interpretation (motion analysis, signal analysis);
- economic ratio interpretation (decision-making analysis);
- stock exchange games (decision-making analysis);
- weather forecasting (forecast and decision-making analysis);
- manufacturing process improvement (industrial robots);
- logistics;
- transport (motion analysis, queue analysis);
- automatic improvement of mental processes (humanoid robots);
- attempts to build an artificial brain (cognitive robots);
- cinematography (avatars, intelligent robots, speech analysis, speaker analysis, identity analysis, biometric analysis, sound analysis, motion analysis).

These examples undoubtedly include James Anderson's work to build a system resembling the structure of the human brain [2], [3]. This project calls for completing three basic tasks within the following areas: preliminary hardware design, programming technologies and software applications. The part proposed in the project and related to selecting and building the hardware architecture is based on the structure of the neocortex in mammals, which can be compared to the 2D connection of elements within the CPU and at the same time associated with the system memory. In order to introduce cognitive software into the above project, it is necessary to have new programming technologies based on topographic data representations, data transmission to the outside, data acquisition from the outside and the use of connection modules to complete the appropriate calculations. Software applications require the simultaneous joining of various elements, which include:

- the natural language;
- programming languages;
- cognitive data analysis;
- information processes taking place;
- decision-making; and
- knowledge management.

Another example of using cognitive methods in computer science is described by Jean-Claude Latombe in his work [59]. He presents the use of the traffic process of automatic robots illustrating this with the example of a new traffic planning approach utilising the plan of a probabilistic traffic map.

Examples of using cognitive methods in computer science are very often connected with data analysis, and in particular image analysis. Such work was conducted in the field of analyses of medical data showing various types of lesions found within an examined organ [116], [117].

Other work in cognitive computer science concerns face, speech, speaker, language and movement analyses. Some of this can be found in the following publications [145], [148], [153], [159].

This kind of work was also conducted to build intelligent mobile robots and unmanned vehicles [1], [73], as well as to adapt psychological and philosophical methods to computer science, develop computational intelligence, create intelligent IT systems, for automatic and machine learning and signal analysis [145], [148], [153], [159].

Chapter 2
Fundamentals of Cognitive Informatics

Cognitive informatics (CI) is a concept which combines the subjects of both the cognitive science and informatics (computer science) based on information mechanisms and processes taking place in the human brain. So cognitive informatics uses natural intelligence merged with engineering applications in interdisciplinary research and science. It covers the use of mathematical theories and descriptions to describe and analyse data and information presented in the form of broad knowledge bases, as well as engineering disciplines including computer science, cognitive science, neuropsychology, system science, cybernetics, computer engineering, knowledge engineering as well as computational engineering.

Theoretical foundations of cognitive informatics are strictly associated with mathematical notions, and the formal models describing theories of cognitive informatics are based on computational intelligence, both of humans and machines. Cognitive informatics can be applied in such areas as so-called cognitive computers, cognitive knowledge bases, cognitive simulations of human brain function, autonomic agent systems, cognitive robots, avatars and computational intelligence.

The term 'cognitive informatics' originated with the beginnings of attempts to carry out interdisciplinary research at the border of informatics, cybernetics, cognitive science, neuropsychology, knowledge engineering, computational intelligence and sciences dealing with human life. Such a combination of research was proposed by authors of the notion of cognitive informatics in their works [142-148].

All research problems coming up in the area of cognitive informatics initially aim at understanding the operating mechanisms of human intelligence and cognitive processes taking place in a person's brain. The understanding of the operation of the above mechanisms is then used to solve engineering problems.

The subjects of IT theories, information theories, various ways of acquiring information, of perceiving and analysing it, and of obtaining objects necessary for the analysis processes have been researched since the time when the information theory was introduced by contemporary informatics until today, when cognitive informatics is developing.

The beginnings of cognitive informatics were associated with the introduction of what is called the contemporary informatics, which studies information as an element distinguished (acquired) from the reality around us, which can be either a representative of a given group of information, or a completely new exception, unknown and never analysed before. Such a perspective on information has shown that processes of information analysis and interpretation are not simple and obvious. On the contrary, this analysis and interpretation should be based on some of the most difficult processes currently known to science world-wide. These are processes taking place in the human brain, because they allow us to understand the essence of the process of reasoning using the information possessed and also lead to the correct analysis of that information.

L. Ogiela and M.R. Ogiela: Advances in Cognitive Information Systems, COSMOS 17, pp. 19–49.
springerlink.com   © Springer-Verlag Berlin Heidelberg 2012

Such a presentation of cognitive information theory allowed it to be defined in a way which unambiguously specifies the place of cognitive information theory among scientific disciplines [142], [144].

Cognitive informatics is understood as the combination of cognitive sciences and informatics intended to study the mechanisms by which information processes operate in the human mind, which processes are treated as elements of natural intelligence and applied to engineering and technical tasks using an interdisciplinary approach.

The above definition of cognitive informatics allows the problems with which this science deals to be identified accurately and at the same time allows the directions of its development to be charted. The most important ones are the development of cognitive IT systems, cognitive robots, avatars, implementing the analysis of human experiences and impressions in automatic (artificial) counterparts, i.e. artificial brains that can perceive, receive and analyse experiences and impressions.

Theoretical aspects of cognitive informatics are considered within two categories [89]:

- the first is the application perspective of the computer science, computer techniques and cognitive research problems like: memory, learning, reasoning, drawing conclusions, analysis;
- the second category is the use of cognitive theories to solve problems in computer science, knowledge engineering, software engineering and computational intelligence.

These problems can be solved by attempting to apply the theoretical foundations of processes taking place in the human brain, such as information acquisition, information representation selection, memory, lost information recovery, communication generation and its process.

2.1 Formal Cognitive Models

What is particularly important in IT systems used in cognitive informatics is their correctly defined system architecture, which in the case of designing theoretical networks of cognitive informatics is based on the models described by Wang. The latter include [142]:

1. **The Information-Matter-Energy (IME) model** - in which information is recognised as one of three complementary components of the natural world, the other two being the matter and energy that surround us. This combination became possible when it was recognised that apart from the basic elements making up the contemporary world - matter and energy - information is treated as its third main component, because the processes running in the human brain are founded on basic functions associated with none else than information processes.

Such a model perspective of the cognitive approach means that, in this model, the natural world (NW) is divided into the following components:

- the world described by physics, the world of the tangible - the physical world (PW);
- and the world of the abstract, the world of the perception - the abstract world (AW), in which both matter and energy are used to describe information (Fig. 2.1.).

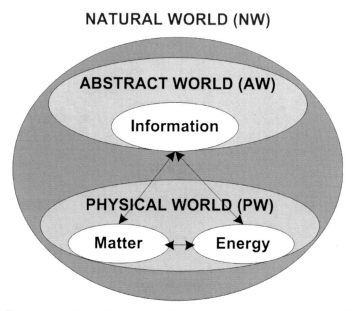

Fig. 2.1. The IME cognitive information model. Source: developed on the basis of [144]

The essence of the presented cognitive information model is to show information as a binder, an important link between two worlds of event description. It is a kind of process of recognising the relations between the biological world operating within the realm of the world of physics, and knowledge in the form of accumulated bases of information taken from the abstract world. The natural world, described by information, matter and energy, is inseparably connected to the abstract world by information learned and perceived using individual, frequently unique techniques dependent on the perception methods and the mental context of the person identifying the specific information. So in the IME model, the definition of information depends on individual methods of human perception, which also serve as the element linking the physical and the abstract world, and which exert a significant impact on the creation of knowledge bases used by cognitive information systems.

As research on using the IME model for the cognitive analysis of data constantly progresses, an improved model which contains one more element - intelligence - has been created.

2. **The Information-Matter-Energy + Intelligence (IME+I) model** - adding the component called intelligence to the IME model represents an extremely important stage in work to build a cognitive information model. The perceived physical world and the invisible world of the abstract are linked not just by information, but also by intelligence which lies at the border of these two completely different realms. This is because intelligence is described as an element of human nature, but is also a notion defined in a very concrete way, so it is a notion coming from the physical world. The difficulty in unambiguously defining and placing intelligence in the natural world is a problem that cannot be solved as there is ambiguity in defining the areas of the abstract and the physical world.

The analysed IME+I cognitive information model (presented in Figure 2.2) indicates the phenomenon of two-stage data analysis using the available intelligence, both human and automatic, the latter founded only and exclusively on structures of human intelligence models.

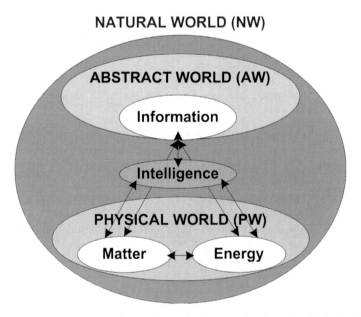

Fig. 2.2. The IME+I cognitive information model. Source: developed on the basis of [146]

Intelligence, presented in Figure 2.2., is understood as the ability of humans or systems to interpret and analyse information and to automatically combine information elements coming from the abstract world, such as: data, information, knowledge, experience. It serves as a connection through which data is transmitted and transferred between information, matter and energy.

3. **The Layered Reference Model of the Brain (LRMB)** – makes use of basic
   and fundamental mechanisms that occur in the human brain and of processes of
   natural, human intelligence. Because processes running in the human brain
   which are referred to as cognitive processes and are therefore of interest to the
   cognitive informatics, psychology, cognitive sciences, neural sciences and neuropsychology are greatly varied, the cognitive analysis process requires that all
   the cognitive processes occur and also that a network is created within which
   all the aforementioned cognitive processes are included. The proposed LRMB
   model describes thirty-nine cognitive processes split into seven main elements,
   namely (Figure 2.3):

- sensations, feelings, impressions;
- memory;
- perception;
- action, process;
- meta-cognitive;
- meta-reasoning;
- highest cognitive levels.

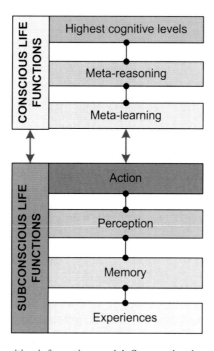

Fig. 2.3. The LRMB cognitive information model. Source: developed on the basis of [142]

The layers of the LRMB system presented in Figure 2.3 are mutually dependent
and complement one another. Each one of them, as a group of life functions,
forms a component of the entire system regardless of whether those functions are
conscious or subconscious.

The operation of cognitive data analysis processes is based on initiating cognitive processes as well as other human life functions, which serve as the foundation for building an intelligent information system.

4. **The Object-Attribute-Relation (OAR) model** – combines methods of representation, recording and analysis which take place in the human brain. It describes layers of the human memory with particular emphasis on the long-term kind, as well as the layer of memory at which a dependency may appear between various metaphors introduced. The OAR model demonstrates relationships between human memory and knowledge using synaptic and neural junctions. This model is used to describe mechanisms and processes taking place in cognitive systems based on the occurrence of the above cognitive processes (Fig. 2.4).

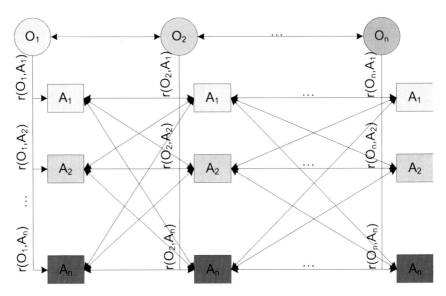

Fig. 2.4. The OAR cognitive information model. Source: developed on the basis of [142]

The operation of the OAR model is made possible by Wang defining the notion of neural informatics (NeI).

5. **The Neural Informatics model (NeI)** – understood as the new, biological/psychological, interdisciplinary perspective on knowledge and information representation taking place in the human brain at the neural level, illustrated using mathematical models. Neural informatics is a branch of cognitive informatics which recognises human memory as the basis and foundation of intelligence, whether natural or artificial.

6. **The model of information representation in the human brain** – this model attempts to demonstrate that any information reaching the human brain is stored in the form of a certain representation which unambiguously belongs to that information. Representations of this type are individual (for every person,

so they should also be so for the computer systems created) and unique (because every person assigns different representations to various information in their brain) - Fig. 2.5.

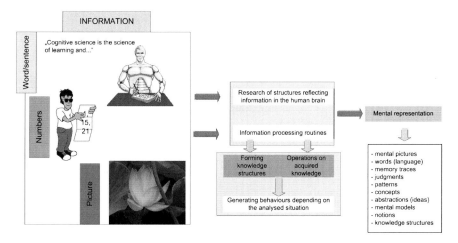

Fig. 2.5. Cognitive information representation model

Information in any form (e.g. in the form of sentences, words, whether spoken or written, sequences of symbols, images, situations, events etc.) reaching the human brain undergoes an analysis its structure, which is to present or reflect this information in the human brain. At the same time, it is processed based on the knowledge obtained earlier by the person (this is the acquired knowledge). Information processing routines produce knowledge structures, which, when combined with operations conducted using acquired knowledge, generate certain behaviours. These behaviours depend on the situation analysed and the set of behaviours generated becomes the basis for storing the information in the shape of a selected mental representation that can take diverse forms. Mental representations created in the human brain take the forms of mental images, words representing a given language, memory traces, judgements, patterns, concepts and ideas, mental models, notions and knowledge structures.

7. **The model based on the structure of the natural mind** – based on the similarities observed between the structure of the natural mind and its corresponding artificial 'imitation'. In this model, elements of the human mind used for cognitive tasks are related to their counterparts (artificial minds) built by cognitive informatics. Perception, memory types, thought processes, data/information analysis and interpretation processes as well as reasoning processes are compared to components of computers, robots, or more generally any type of a machine. For a human, the design of a cognitive model resembles the structure of a block cognitive/decision-making system, whereas for a computer, the structure of its artificial counterpart.

8. **The natural intelligence model** – is a type of system solution combining known theories of intelligence, which include the structural theory of intelligence, the hierarchical theory, the development theory and concepts of activities aimed at improving the capabilities of an individual, the assimilability and understanding of new experience. In addition, this model shows that (both for humans and machines) solving new problems and dealing with new situations requires activating the information processing system in which the processing, analysis, interpretation and reasoning processes run in a controlled and conscious way. In addition, they are complemented by elements of human intelligence, whose domain covers executive processes, i.e. planning, controlling, adjusting, the processes of stimulating and controlling the strategies of action selected.

9. **The model of human perception processes** – based on processes of interpreting sensory data using contextual aspects, a certain attitude and knowledge resources aimed at the correct recognition of the object. The essence of this approach consists in description processes and research processes showing the given object/information/data as the foundation of the perception process, which is subject to a complete identification. The process of identification (recognition, understanding) does not yet lead to complete analysis, because it is only the identification and recognition of factors/processes impacting the course/change of the analysed situation that supports the claim that the entire process was successful.

10. **Cognitive computer models** – a combination of three main components including artificial intelligence, cognitive science as well as system applications and solutions (Fig. 2.6). Due to the connections and implications between these components, they form solutions called cognitive computers. Models of cognitive computers are created by combining cognitive solutions from the field of artificial intelligence in which natural intelligence processes adopted for artificial analysis purposes are identified and cognitive processes which form the basis for problems of the semantic analysis of data. System applications are developed using the solutions adopted in the artificial intelligence field and supplementing them with cognitive solutions. In addition, cognitive computer models activate natural intelligence processes to distinguish the correct cognitive solutions and structures. Cognitive analysis processes executed by cognitive computers are based on semantic (cognitive) solutions which become the foundation of the applications developed. In addition, artificial intelligence solutions are of great importance for designing software, which represents a significant component of the applications created.

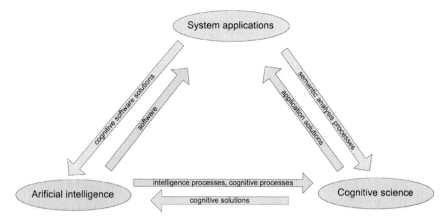

Fig. 2.6. Cognitive computer models

11. **Cognitive machine models** – just as the cognitive computer model, this is the combination of three main components, namely artificial intelligence, cognitive science as well as system applications and solutions (Fig. 2.7). Due to the connections and implications between these components, they form solutions called cognitive machines (robots used for diverse jobs). These may be production machines, but they may also be humanoid robots or artificial brains. Cognitive machine models are created by stimulating cognitive solutions applied to artificial intelligence tasks and activating processes of natural intelligence to identify the appropriate cognitive solutions and structures. In addition, cognitive analysis processes are applied to semantic (cognitive) solutions which become the foundation of the application solutions developed. Artificial intelligence solutions are also of great importance for developing control mechanisms, which represent a significant component of the applications developed.

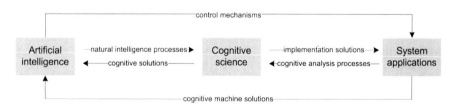

Fig. 2.7. The cognitive machine model

12. **The Cognitive Model of Memory** (CMM) – is a model operating based on, and by the interconnection of various models of human memory, including:

- The SBM (Sensory Buffer Memory) model – a memory model characteristic for connections of the memory to senses.

- In this model, every stimulus received is kept for a certain (usually very short) time in the form of an individual analogue code. At the same time, every stimulus is processed by the cognitive system. The entire analysed information on the stimulus situation received goes to a so-called sensory store of a very large capacity. The operation of the SBM model is automatic, principally devoid of a control process.
- The STM (Short-Term Memory) model – a memory model in which various information is stored, particularly that coming from the SBM model storage. The STM model is made up of:

  - information obtained from the LTM model;
  - sensory stimulation processes;
  - current information processing.

The fundamental function of the STM model is assumed to consist in storing information over the time necessary for its processing. The processing stage is understood as an analysis consistent with the goal set and chosen when executing a given process, e.g. finding a piece of paper and a pen in order to jot down some important detail. This is why the STM model is characterised by a very short duration of information storage.

- The LTM (Long-Term Memory) model – a model of memory which, in contrast to the STM model, is characterised by a long duration of information storage. This model stores various types of knowledge and provides the diversification and flexibility allowing the individual to adapt to the environment. This process is possible, among other reasons, due to the semantic knowledge collected in the model.
- The ABM (Action-Buffer Memory) model – a model on the border of statics and dynamics associated with memory processes. It can be activated in layers of the operating memory. This memory is subject to two types of limitations associated with its capacity. The first is the statics of the information storage processes (the buffer), and the second is the dynamics of information stimulation (activation) processes. In this model, just a few elements can be activated, and as a result of this process they become elements consciously processed in memory. The activated elements become the focus of attention and concentration, which means that the remaining elements are then inaccessible and cannot be analysed and processed. This model, by activating a certain group (number) of long-term memory elements, at the same time restricts access to the remaining group of memory process elements (active at the previous operation of information processing).
- The semantic memory model – mainly shows the stages of information processing based on the knowledge of verbal symbols, as well as their role and significance in analysis processes. Verbal symbols are obtained from layers of natural language specifying the relations between specific symbols, the rules of symbol analysis and description, relations, concepts, formulas as well as algorithms for modifying and processing given symbols. Resources of semantic memory are

created based on the knowledge collected about the significance of verbal symbols as well as the relations and dependencies between them.

13. **The cognitive model of awareness and machine cognition** – a model of cognitive informatics founded on a model of human awareness and human cognition transferred into the field of machine execution of the above cognitive tasks. The analysis of cognitive processes, including the analysis of consciousness, dates back to the times of Aristotle, when work was begun to analyse the human body and brain [27], [56], [62], [132], [152]. Consciousness was treated as the basic element describing and characterising the life and mind of humans, so in cognitive informatics publications is has been assigned features connected not only with human cognitive aspects, but also machine elements. So consciousness is understood as one of life functions that works within the LRMB model (described previously) and represents conscious and subconscious life functions.

The presented model of consciousness and machine cognition uses the learning theory of both the human and the machine learning process. This is because learning is a cognitive process of acquiring, obtaining and transferring knowledge and behavioural experience. The psychological variety of learning processes is impressive and has been described at great length in the literature of the subject [62], [131], [133].

In cognitive science, learning is synonymous with the appearance of relatively permanent changes in the cognitive and thinking processes as well as in perception, understood as the results of certain impressions and experience gained during the entire learning process [29], [41], [120], [122], [131], [152].

Cognitive informatics borrowed certain items from the cognitive learning process characteristic for the human mental functions to adapt learning processes and make them executable by machines. This is why cognitive informatics defines machine learning processes as cognitive processes taking place at the level of cognitive meta-processes described for the LRMB model. These processes influence one another as part of the following tasks:

- object identification;
- abstracting;
- searching;
- introducing a concept (defining);
- understanding;
- comprehending;
- remembering; and
- retrieving and testing memory.

The most important element of the learning process is memory, in particular its long-term kind (the LTM cognitive model). All machine learning processes found in cognitive informatics are based on the LTM and OAR cognitive models, where simulations of the operation of natural minds form the foundation for creating all the possible relationships occurring in human minds and then adapting them to machine learning processes.

Memory models are also distinguished by the methods of remembering, storing and retrieving information from various layers of memory. Differences of this type become the foundation for establishing work stages in the operations of cognitive data analysis models. These operations are applied in automated (computerised) solutions in which the stages of information sourcing, its analysis, recording, storage and retrieval are distinguished within the work of the system.

Models of human memory have diverged somewhat, because in 1984 Endel Tulving distinguished five basic criteria aimed at demonstrating the distinction between particular memory systems. These criteria have the form of the following rules [138]:

- memory systems operate according to different rules;
- they are characteristic for different stages of the ontogenetic and phylogenic development;
- memory systems have various neural foundations;
- they are characterised by different behavioural and cognitive functions, they are also used to process various types of information;
- information is represented using various formats.

The presented memory models show memory as a large set, a store within which various tasks are executed. This view of memory is characteristic for block models of memory likened to computer systems which are modular and whose action is sequential. The best known block memory models include the one proposed in 1968 [5] by Atkinson and Shiffrin (Figure 2.8).

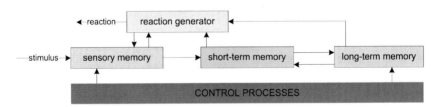

Fig. 2.8. Block memory model. Source: developed on the basis of [5]

Figure 2.8. shows a block model of memory consisting of three memory types – SBM, STM, LTM – within which the stimulus reaching the memory is stored. Information can reach every block independently and in the case of sensory memory, the STM and LTM memory can be bypassed. The block memory model behaves similarly in the case of the short-term memory, within which the recorded stimuli fade away. In the case of the long-term memory, the reaction to the stimulus may occur directly or through the short-term memory [67], [75].

In 1988, Lehrl and Fischer [63] proposed a completely different block model of memory representation (Figure 2.9).

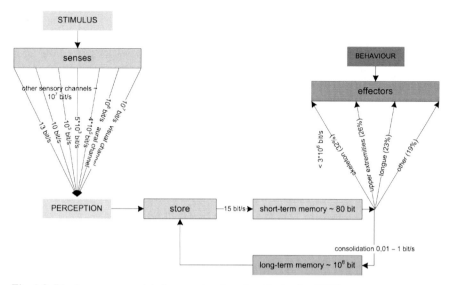

Fig. 2.9. Block memory model. Source: developed on the basis of [63]

Another feature of the memory model presented here is that information processing operations are measured in seconds and bits. The seconds tell how information processing operations change over time, and the bits represent the quantity of information processed. The authors of this concept found that the human memory system is capable of processing between 10^9 and 10^{11} bits of information per second. Stimuli reaching human senses through all sensory channels, within which some 10^7 bit/s are processed, are sent to a kind of store, directly connected to memory. The stimuli sent to short-term memory fade away or end up in long-term memory, from which they can return to the store.

The results of the operation of the above models can be seen in applications of cognitive informatics and neural informatics models, in which the LTM (long-term memory) model plays an important role:

- using the LTM model for subconscious processes;
- the operation of long-term memory during sleep;
- the operating mechanism of the LTM model during sleep;
- the operation of the LTM model with reference to the OAR model;
- the role of the visual function and the perception function (including daydream creation) in the LTM model.

The above formal models of the cognitive approach to data analysis and interpretation are the result of work conducted to build the correct system architecture presented in cognitive informatics subjects.

2.2 Cognitive Resonance Model

An information system which analyses data based on the individual features
characteristic for the specific type of data holds in its base the indispensable
knowledge, which, during the analysis processes conducted, becomes the basis for
generating the system's expectations. These expectations are generated automatically using the expert knowledge bases collected in the system. At the same time,
the information system executes an analysis aimed at identifying and indicating
features characteristic for the analysed data sets. As a result of combining significant characteristic features of the analysed data with expectations generated using
the knowledge held in the system concerning the analysed semantic content,
cognitive resonance occurs (Figure 2.10.) [87], [97], [114].

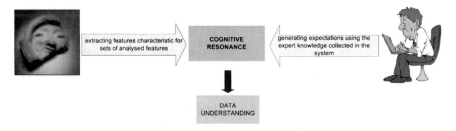

Fig. 2.10. The cognitive resonance phenomenon

Cognitive resonance becomes the cornerstone of the process of understanding
(as such) the analysed data sets. This understanding occurs as follows. The stream
of expectations generated by certain hypothetic semantic content (meaning) of
data and the stream of features characteristic for the specific data set are compared, as a result of which certain pairs of expectations and features identified in
the analysed data can gain in importance (become significant) or, conversely, lose
importance (become insignificant). This comparison, by causing cognitive resonance, leads to confirming one of the possible hypotheses (in the case of data
whose content can be understood), or conversely, shows that the inconsistency of
features and expectations cannot be eliminated. The first case means that the
analysis process conducted was successful, the second means that the attempt to
automatically understand data failed [92], [97].

Information systems using cognitive resonance for data analysis are based on
methods defining structural reasoning techniques for matching patterns [89], [92].
The structure of the image being analysed is compared to the structure of data
constituting the pattern during the analysis process. The comparison is made possible by using strings of derivation rules which enable the pattern to be generated
unambiguously. These rules, referred to as productions, are defined in a specially
introduced grammar, which in turn defines a certain formal language. Data thus
recognised is assigned to the class to which the pattern representing it belongs.
The cognitive analysis utilised in cognitive information systems very frequently

uses a syntactic approach which employs functional blocks for the semantic analysis and interpretation of the image [92]. In this case, the input image undergoes pre-processing, which consists of:

- filtering and pre-processing the input image;
- approximating the shapes or locations of the analysed objects; and also
- coding the image with terminal components of the introduced language.

The completion of these stages represents the data anew in the form of hierarchical structures of a semantic tree and produces subsequent steps of deriving this representation from the initial symbol of the grammar [114], [117].

While pre-processing image data, a cognitive recognition system must in most cases execute the segmentation, identify picture primitives and also determine the relations between them. The classification proper consists in recognising whether the given representation of input data belongs to the class of data generated by the formal language defined by one of the grammars that can be introduced. Such grammars can be classified as sequential, tree and graph grammars and recognition with their use takes place during a syntactic analysis conducted by the system [36], [92], [97], [114].

Traditional data analysis processes based on cognitive resonance have been extended to include stages at which the system learns using the knowledge collected in its knowledge bases and by analysing situations it does not understand. In this case, if the system encounters a situation it does not understand, i.e. one undefined in its knowledge base so far, it cannot correctly classify it and match the pattern. Then the system enters a state of 'surprise' and incomprehension of the analysed data, which state it can, however, use to supplement the knowledge base with new cases of pattern classification and data understanding, as a result of which it becomes necessary to add new, undefined examples to the expert knowledge base. The process by which the system learns is presented in Figure 2.11.

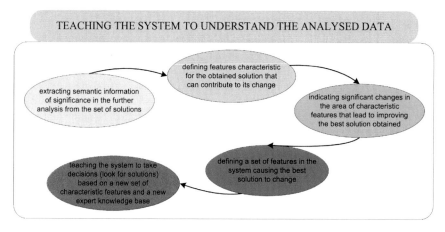

Fig. 2.11. System training process

In data analysis processes run by cognitive information systems, a certain feature characteristic for these very systems is noted. Namely, data analysis processes cannot end at the stage of recognising the analysed data sets, because the essence of the process conducted is to extract the features characteristic for this data and determine their semantics. This is why the processes of automatic data analysis now include a stage of automatic understanding executed using artificial intelligence technologies, which, apart from the simple recognition of the item to be analysed, can also extract significant semantic information from the set. This information allows its meaning to be interpreted, in other words, supports its full understanding.

The system learning process progresses in five stages. At the first stage, the semantic information which can be of significance in the further analysis process is extracted from sets of solutions, so if some information is useless in the current data analysis process, it is treated as superfluous for that process and will be omitted from it. After this stage, the features characteristic for the solution obtained at the first stage of the analysis process are identified. Identifying that type of features can cause a change of the solution obtained at the first stage, e.g. because the set of expectations or the expert knowledge base is extended and thus new patterns are defined. This moment marks the beginning of the following stage which focuses on indicating significant changes in the field of characteristic features leading to improving (optimising) the solution formulated, as a result of which, a redefinition is carried out in the set of characteristic features in the system. The last stage of system learning is that of looking for solutions based on a new set of characteristic features and a new expert knowledge base, which at this stage takes the shape of a set of new patterns defined in the cognitive system.

Supplementing data analysis with stages at which the system learns new solutions means that the cognitive resonance must be repeated in the data analysis process, and if the learning process is multiplied, then cognitive resonance must be repeated more than once. Incorporating new system learning solutions in the data analysis process makes those data analysis processes much more complex (Fig. 2.12).

After the stage at which cognitive systems learn new solutions as part of data analysis processes, the analysis and understanding process based on cognitive resonance is repeated, but unlike the traditional analysis process, is now makes use of new (extended) sets of analysed data and a new (extended) base of expert knowledge. It is these very elements that become the primary foundation of cognitive data analysis processes in new analysis and understanding systems enhanced with aspects of cognitive system learning.

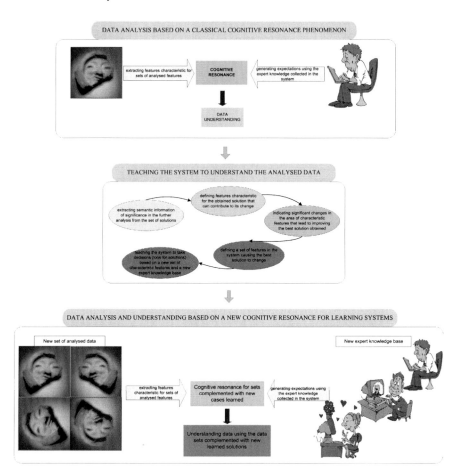

Fig. 2.12. Cognitive resonance in the data analysis and understanding by learning systems

Processes of the cognitive analysis of data are based on data categorisation, which can be split into two fundamentally different categorisation methods: semantic categorisation and syntactic categorisation.

2.3 Semantic Analysis

In semantic analysis processes, a major role is played by processes modelled on human cognitive/decision-making processes, particularly those consisting in the description, analysis and interpretation of the meaning contained in analysed data sets. These processes are inseparably associated with cognitive analysis, during which a significant role is played by semantic analysis stages.

Semantic analysis processes are used for designing automatic data analysis systems, among which cognitive systems represent an important class.

Semantic analysis is the key to the correct operation of cognitive data analysis systems. When this analysis is conducted, several different (but equally important for the analysis) processes occur: the interpretation, the description, the analysis and reasoning.

The main stages of semantic analysis are as follows:

- data pre-processing:

    - filtration and segmentation;
    - approximation;
    - coding;

- data representation:

    - recognising picture primitives;
    - identifying relations between picture primitives;
    - defining relations between objects in the image;

- linguistic perception;
- syntactic analysis;
- pattern classification;
- data classification;
- feedback;
- cognitive resonance;
- data understanding.

The main stages of semantic analysis are shown in Figure 2.13.

A clear majority of the above semantic analysis stages deals with the data understanding process, since beginning with the syntactic analysis conducted using the formal grammar defined in the system, there are stages aimed at identifying the analysed data with particular focus on its semantics (the meaning it contains). The stages of recognition itself become the starting point for further stages, referred to as the cognitive analysis. This is why the understanding process as such requires the application of feedback during which the features of the analysed data are compared to expectations which the system has generated from its expert knowledge base. This feedback is called the cognitive resonance. It identifies those feedbacks which turn out to be material for the analysis conducted, i.e. in which features are consistent with expectations. The next element necessary is the data understanding as such, during which the significance of the analysed changes for their further growth or atrophy (as in lesions visible in medical diagnostic images) is determined.

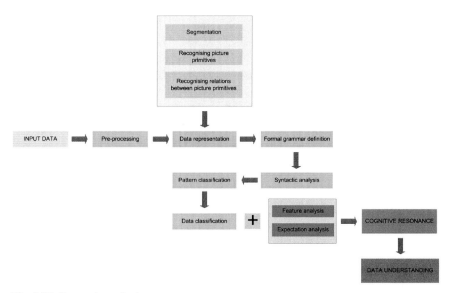

Fig. 2.13. Semantic analysis stages

Data is analysed by identifying the characteristic features of the given data set, which then determine the decision. This decision is the result of the completed data analysis (Fig. 2.14).

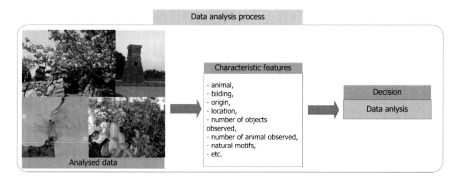

Fig. 2.14. Data analysis process

The data analysis process is complemented with the cognitive analysis, which consists in selecting consistent pairs and non-consistent pairs of elements from the generated set of features characteristic for the analysed set and from the set of expectations as to the analysed data generated using the expert knowledge base kept in the system. The comparison leads to cognitive resonance aimed at identifying consistent pairs and non-consistent pairs, where the latter are immaterial in the further analysis. In cognitive analysis, the consistent pairs are used to understand the meaning (semantics) of the analysed data sets (Fig. 2.15).

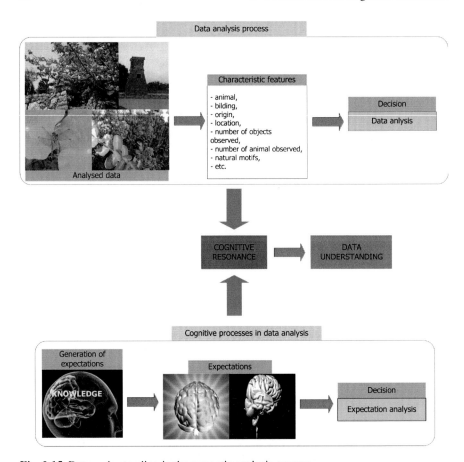

Fig. 2.15. Data understanding in the semantic analysis process

Because of the method of conducting semantic analysis and linguistic perception algorithms - grammar formalisms - used in its course, semantic analysis has become the core of the operation of cognitive data analysis, interpretation and reasoning systems.

Semantic analysis processes which form the cornerstone of cognitive information systems also underpin a new branch of science, which is now developing very fast: cognitive informatics. The notion of cognitive informatics has been proposed in publications [142], [143] and has become the starting point for a formal approach to interdisciplinary considerations of running semantic analyses in cognitive areas.

Cognitive informatics is understood as the combination of cognitive science and information science with the goal of researching mechanisms by which information processes run in human minds. These processes are treated as elements of natural intelligence, and they are mainly applied to engineering and technical problems in an interdisciplinary approach.

Semantic analysis in the sense of cognitive analysis plays a significant role, as it identifies the meaning in areas analysed. The meaning as such is identified using the formal grammar defined in the system and its related set of productions, within which productions are defined, which elements the system utilises to analyse the meaning. Analysis processes are applied to features such as (in the case of diagnostic image interpretations):

- the shape, the present feature, lesion, pathology;
- ratio;
- lesion occurrence;
- its size, length, width;
- lesion vastness;
- number of lesions observed;
- number of repetitions of the given situation, lesion, pathology;
- lesion structure;
- pathology location.

These features can be identified correctly using the set of productions of the linguistic reasoning algorithm. For this reason, the definition of linguistic algorithms of perception and reasoning forms the key stage in building a cognitive system.

In line with the discussed cognitive approach, the entire process of linguistic data perception and understanding hinges on a grammatical analysis aimed at answering the question whether the data set is semantically correct from the perspective of the grammar defined in the system, or is not. If there are consistencies, the system runs an analysis to identify consistencies and assign them the correct names. If the is no consistency, the system will not execute further analysis stages as the lack of consistency may be due to various reasons. The most frequent ones include:

- the wrong definition of the formal grammar;
- no definition of the appropriate semantic reference;
- an incompletely defined pattern;
- a wrongly defined pattern;
- a representative from outside the recognisable data class accepted for analysing.

All these reasons may cause a failure at the stage of determining the semantic consistency of the analysed specimen and the formal language adopted for the analysis. In this case, the whole definition process should be reconsidered, as the error may have occurred at any stage of it.

Cognitive systems carry out the correct semantic analysis by applying a linguistic approach developed by an analogy to cognitive/decision-making processes taking place in the human brain. These processes are aimed at the in-depth analysis of various data sets. Their correct course implies that the human cognitive system is successful. The system thus becomes the foundation for designing cognitive data analysis systems.

Cognitive systems designed for semantic data analysis used in the cognitive informatics field execute data analysis in three 'stages'. This split is presented in Figure 2.16.

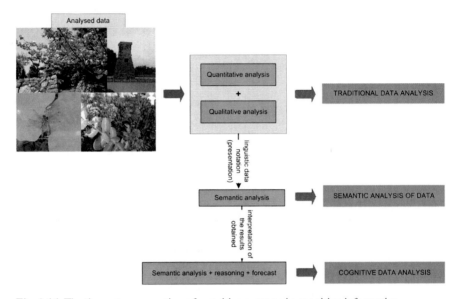

Fig. 2.16. The three-stage operation of cognitive systems in cognitive informatics

The above diagram shows three separate data analysis stages. The first one is the traditional data analysis, which includes qualitative and quantitative analyses. The results of this analysis are supplemented with the linguistic presentation of the analysed data set, which forms the basis for extracting semantic features from those sets. Extracting the meaning of the sets from them is the starting point for the second stage of the analysis, referred to as the semantic analysis. It is conducted according to a definition of the linguistic data perception and analysis, using the grammar formalisms introduced. The end of this stage at the same time forms the beginning of the next analysis process, referred to as the cognitive data analysis. During this stage, the results obtained are interpreted using the semantic data notations generated previously. The interpretation of results is not just their simple description or a recognition of the situation being analysed, but it is particularly a process in which data, the situation and information are understood, the stage of reasoning based on the results obtained and forecasting the changes that may appear in the future.

2.4 Semantic Categorisation

Semantic categorisation is important for the analysis of the content and meaning present in data sets. The most widely used type of semantic analysis is lexical analysis, in which the structure of the word is used to describe it together with the content this word carries. This is why the correct assignment of the word to the language, which becomes the basis and the pattern of the analysis conducted is extremely important in the entire analysis process.

In semantic analysis processes, an important role is played by the time during which it is checked whether the given case is a representative (a member) of a specific category, or not. The meaning of the lexical concept is represented by specific semantic features which play a primary role or define the meaning of the word (Figure 2.17).

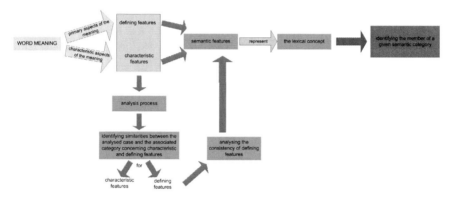

Fig. 2.17. The process of identifying a member of a given category

The meaning of the analysed word, which in the semantic analysis may relate to the description of basic aspects of the meaning of this word constituting the set of defining features making it possible to identify the semantic features, also helps to extract the characteristic aspects of the meaning of the analysed word. These aspects are called characteristic features, and just as defining features, they contribute to identifying the semantic features of the word. Both the characteristic and the defining features become the foundation for conducting the analysis process which leads to identifying similarities between the analysed word and the category to which it belongs. At this stage, similarities are found between the characteristic aspects of the meaning of the given word and the category to which it belongs, and between the primary aspects of the meaning of the analysed word and its associated category. The result of this action is an analysis of the consistency of defining features, which ends in specifying the semantic features of a given word representing lexical concepts, which concepts are in turn used to identify the semantic category to which it belongs.

The language and the forms for describing the meaning of a word are not sufficient to conduct an in-depth analysis because they do not in themselves have a system of features describing the sound aspect of the language. In this regard, we have to refer to the phonological system currently used, which has a rich system of features called distinctive ones, used to describe the sound aspect of the language analysed. For the phonological system, it is important to indicate features which can be phonologically relative if we define the function of the language as the communication between its users. The basic functions that the phonological system plays are presented in Figure 2.18.

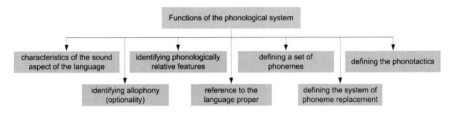

Fig. 2.18. Features of the phonological system

The phonological system depends on the hierarchy of segments found in the analysed languages, and there can be several such hierarchies. The phonological analysis of the language makes use of phonemes and segments which can be distinguished among them, namely phones and sounds. The significance of such segments cannot be overestimated, because placing them in pairs frequently means that we hear similar sound forms, but their meanings are completely different, e.g. pet and bet, cat and get, dill and till, etc. What is extremely important in demonstrating the difference between analysed sounds is showing different phones – distinctive features that differentiate the sounds studied. In our example those are pairs of phones [p] and [b], [c] and [g] as well as [d] and [t]. However, it is worth noting that it does not suffice to just indicate the different phones in the analysed words, because phonological analysis aims at specifying the articulation locations and the acoustics of the analysed features.

The phonological analysis starts with segmentation consisting in splitting the word into the smallest sound units (phones and sounds) characterised by the same features. Then those features among them which occur several times (repeat) as a result of a certain coincidence or being adjacent to the same features are ignored. The next stage is to detect features whose elimination would not cause a change in the meaning of the word, even though they are undoubtedly audible in its pronunciation as such. This process leads to swapping the given features for features characteristic for other words. Then the features that can be eliminated without changing the meaning of the analysed words are detected (revealed). Finding all pairs of words in which the meaning changes as a result of removing similar features leads to creating a whole base of phonological oppositions, in which distinctive features are identified once again. The base of phonological oppositions is to serve in the whole phonological analysis process to find pairs of words for which the phonological oppositions are minimal (the smallest possible) and to build a hierarchy of the studied features.

In phonological analysis it is worth noting that in the whole process of creating phonological oppositions some exceptions can be encountered, namely phones with a complementary distribution which makes the difference between them acoustic and articulatory. In addition, these phones do not occur in the vicinity of one another which means that they are opposed to other phones, but not to one another, so such phones are classified as phones belonging to one phoneme, because their common features do not characterise any other phone.

A matter of utmost importance in phonological analysis and semantic categorisation is the increasingly frequent problem of speech analysis and the analysis of the speaker, i.e. the person pronouncing specific words. This is why the next chapter presents the problems of speech and speaker recognition.

2.4.1 Using Semantic Categorisation to Analyse Speech and the Speaker

As an element of semantic categorisation, the structural approach hinges on the analysis of the input data structure. It has already been said that this chapter will focus on phonological analysis, and thus speech analysis. The entire process of the discussed data analysis ends in a stage referred to as recognition, the stage most important from the point of view of the analysis, because at this stage the question is to be answered whether we have been able to recognise the analysed text at all (spoken or written) and whether it is logical from the perspective of the analysing person.

Two primary subjects of recognition are distinguished in the process of text recognition by its semantic categorisation: speech recognition and speaker recognition.

The first type of recognition, speech recognition, starts at the speech processing stage which is to end in extracting significant features and their meanings used in subsequent analysis. In this case, the analysed speech is equated with the contents of the text, independent of the person saying the specific words. For this reason, during the analysis, the speech signal is identified together with all its elements that can in any way determine the contents of the text.

Speech analysis is conducted very frequently due to rather significant benefits of speech, which we frequently seem to forget, but which play important roles. These benefits definitely include the natural way of transmitting the message, including requests, instructions, requests, prohibitions and spontaneous content. Communication itself is a very important aspect and element of speech, as it is much faster than any written form of text. In addition, it contains some expressiveness, because when we want to say a certain sentence, we often add our own emotions to it and this is sufficient, whereas in a written sentence we need to add the appropriate number of epithets to at least partly convey the emotions that came across with the verbal form. The spoken form frequently also dominates in social, interpersonal interactions, as it is faster and simpler to call another person to convey a message than to write a traditional letter, an e-mail or a text message. This form of conveying information is much more dominant in interpersonal contacts for purely practical reasons, although it has to be noted that the younger generation – the 'computer' or 'mobile' generation – prefers to transmit information over phones or modems. Still, the advantages of speech are incomparably greater than of written text, even if we should not underestimate or in any way belittle the advantages of written text.

Speech recognition itself very frequently boils down to recognising sentences in which the vocabulary, the sequence of words and the dependency on the speaker play important roles. The vocabulary, as the factor determining the selection of the speech recognition method, is split into small or large vocabulary sets. In addition, these sets may be limited or unlimited, which in the case of limited sets may cause wrong recognitions, and in the case of unlimited ones – recognition ambiguity. Such sets also have one more disadvantage, as we should remember that our vocabulary and the sequence of words is interrupted (with pauses) or it has no pauses, and then the places where one word significant for the sentence being recognised ends and the next word begins have to be market independently.

Another important element in recognising speech is its reference to a specific speaker or the lack of such a reference. In this regard, we say that speech is speaker-dependent, or is independent of any speaker. At this point, it is also worth adding that speech (sentences spoken by a specific person) does not always have significant content, and what is more, is frequently corrected or changed by the speaker while speaking. For this reason, speech recognition is much more complicated than recognising written text, which, once written, does not change during the analysis. We are also frequently surprised while speaking by some spontaneity which slips into our sentences, whether we want it or not. This is another element of instability of a statement and its volatility which is evidence to the difficulties in speech recognition.

Speech recognition starts when the sentence is said. This sentence becomes the original data for the analysis process, and then the possibility of certain interference appearing has to be accounted for. This interference includes not just system interference, but also all kinds of interference coming from the environment. At this stage of the process, the recognising system treats the sentence as a noisy sentence within a channel of noise, so the recognising system immediately begins the decoding stage, which will produce a recognised sentence. The entire recognition process resembles what happens in a school hallway, where student A standing at one end of it says sentence x, while student B standing at the other end must understand the sentence spoken to him by student A. During this short discussion between students A and B, loud school din prevails in the hallway. Regardless of the terrible noise, student B may hear sentence x, but he might also only guess the content of the sentence said by student A, and then there are two possible solution of the problem : either he successfully guesses the contents of sentence x, or it will be distorted and student B will recognise completely different words in the sentence and will build sentence y of them. This situation illustrates the speech recognition process and the threats posed by a whole range of system interference. In this case it consisted in the noise, din and the distance between the students, whereas in the system analysis there is no distance difference, but undoubtedly various other sources of interference can appear.

A diagram of the process of recognising speech as a sound signal is presented in Fig. 2.19.

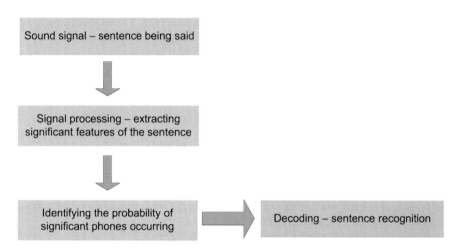

Fig. 2.19. Stages in the speech recognition process

This diagram shows the speech recognition process from the moment the sentence is pronounced and becomes a sound signal for the system until the moment the sentence is recognised. The input sound signal undergoes signal processing to extract significant features (usually acoustic) of the sentence, and then the probability of the occurrence of individual phones of major significance for the analysed sentence is identified. This probability can be determined using probability calculus methods or neural networks. After the probability of occurrence of significant phones has been determined, there comes the stage of information decoding leading to recognising the sentence. This way, the entire speech recognition process is executed.

Another type of recognition analysis is recognising the speaker, i.e. the person pronouncing the given words. In this analysis it is not the content of the statement that is analysed, but only the person making that statement, so the speech signal is analysed to enable distinguishing different speakers from one another not only if they say different statements, but primarily when their statements are similar or identical. This is why the entire speaker recognition process starts by creating the appropriate base of expected speakers from which the analysing system will select the ideal speaker, i.e. the one who ideally corresponds to the system, or will announce the lack of one due to no or too little consistency between the speaker and the speaker patterns kept in the system. If the system has a properly designed database containing a certain (large) number of reference models for which the system will try to identify references, the speaker verification process follows, which consists in the speaker confirming his/her identity. This process stage is of great importance, as without speaker verification, an identification could not be completed even if the person about whom the system is certain that he/she is the speaker searched for is correctly recognised. After the verification stage is completed, the analysing system begins the second stage, referred to as 'speaker identification'. This stage identifies as such which of the speakers known to the system is speaking. Thus this is the stage at which the entire recognition process can finish. If the speaker has been correctly linked to a person included in the system reference set (called the system base), the recognition process can be judged fully successful, but this success is not always possible as the set constituting the database may be too small and may not contain the correct references to the speaker analysed at the given moment. Of course, disruptions may occur in the entire analysis process or the analysing hardware may fail, and as a result the recognition process ends in failure. However, such situations occur relatively rarely and they do not paralyse the analyses conducted to any significant extent.

Even though the speaker analysis process does not lead to analysing the contents of the spoken text, it can depend on the type of text pronounced. In the case of speaker analysis independent of the spoken content we can claim that this analysis should proceed correctly and what is more, should end in success, namely speaker recognition, for any statement of this speaker. Yet it is worth adding that this type of analysis is used in situations in which we are convinced that cooperating with the speaker will not produce any benefits or when the speaker does not want to cooperate with us (as the people conducting the analysis). In other cases, methods of speaker recognition that may depend to some extent on the content said out loud are very often used. These are cases where the speaker is suspected of using or pronouncing a password or a fragment thereof, uses numbers coded in the system or pronounces a text provided by the analysing system. In these cases, the speaker recognition is said to be dependent on the content pronounced by the speaker.

Thus speaker recognition is determined by the presence of a certain type of a link between the specific speaker and a certain pattern defining this speaker. If the system contains a pattern defining the speaker, then in the recognition process the system will assign the speaker to his/her reference model. If the opposite is true, then the system may indicate the reference pattern closest to the speaker, or if the defined number of consistent features is not achieved, it may signal that the recognition analysis failed. The case in which the speaker corresponds ideally to the pattern with which he/she has been associated and to which he/she is assigned is called 'speaker identification from a closed set of speaker models', whereas if a certain number of consistent features is analysed because there is no full compliance, this is referred to as 'identifying the speaker from an open set of speaker models'.

The speaker recognition process, regardless of the type of the speaker and of the analysis conducted, consists of three fundamental stages, namely: the preliminary processing of the speech signal, the selecting features significant for the given speaker including extracting these features from among others that have no major impact on current analysis results, as well as classifying the speaker. Each of these three stages of the entire recognition process is shown in Figure 2.20.

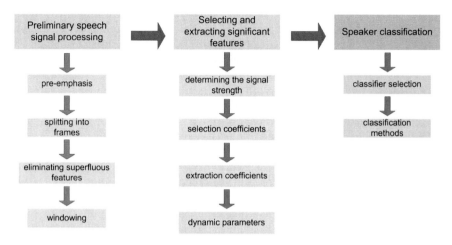

Fig. 2.20. Speaker recognition process

The first stage – speech signal pre-processing – consists of four basic activities and tasks, namely the pre-emphasis, splitting into frames, eliminating superfluous features and windowing. At this stage, two tasks mentioned at the beginning of this chapter are executed: the speaker is verified and identified. The second stage – significant feature selection and extracting – begins with determining the signal strength, choosing selection coefficients, selecting extraction coefficients, and ends in an analysis based on dynamic parameters. The last stage – speaker classification – boils down to selecting the appropriate classification methods and classifiers.

Speech and speaker recognition are complex processes and even if they appear independent, they have certain joint characteristics and feature some common challenges. The latter certainly include developing a comprehensive speech recognition system for our mother tongue, developing a system recognising both continuous speech and separated words, as well as designing integrated speech recognition systems combined with dialogue systems.

Speech analysis and speaker recognition systems use semantic analysis, though this applies not only to systems of this type. It is used with equal success in categorisation processes, e.g. of medical systems.

2.4.2 Semantic Analysis in Medical Systems for Cognitive Data Interpretation

Semantic categorisation need not always apply to speech, as there are classes of topics which, when subjected to semantic analyses, splendidly illustrate its cognitive potential. The authors have spent several years researching the use of semantic analysis for the cognitive categorisation of, *inter alia*, images of the central nervous system including changes in the appearance of the spinal cord, lesions of foot bones and of palm bones. The purpose of the medical image analysis conducted was to find what deformations have occurred in the examined organ and what disease processes they may be symptoms of. Just recognising the analysed shape and an attempt to classify it are not enough, as these can very easily be disturbed, even if only by the natural differences between individuals concerning the structure of the organs examined, and also if there is no lesion present. With regards to cases of an organ in which there actually is a lesion, the analysis conducted is aimed at confirming the occurring pathology of this organ and identifying other information significant for both the physician and the patient.

The semantic analysis of medical images, just as speech analysis, suffers from some difficulties, one of which is certainly the lack of a universal pattern of an ideally healthy human organ due to the varied anatomical structure of the organ in various people. Creating a database containing a certain number of perfectly defined patterns does not mean that all possible situations (pathological cases) have been classified and their representations included in the system base. Consequently, one has to account for the fact that the same disorders cause deformations differing in size and shape and in addition occurring in different places within the examined organ, and for situations in which (apparently) very similar changes in the shape of the organ image are associated with completely different clinical interpretations.

Due to the possible difficulties that can occur in the diagnostic interpretation of each analysed medical image, it can justly be said that the recognition itself of lesions in the analysed image is not sufficient to attempt a diagnosis, whereas attempting to understand the image in the sense of determining the nature of the process that led to this and not another deformation of the organs visible in the image offers much better opportunities for an accurate interpretation. This is why it makes sense to make the automatic data analysis process go beyond a superficial analysis of the image form and an attempt to classify it, and instead attempt to understand the image. Machine understanding of the analysed medical image can be used as an example of a task requiring in-depth, cognitive data analysis. The entire process leading to understanding the image is extremely complex and its individual phases are inseparable from and interdependent on one another. This process is illustrated in Figure 2.21, beginning with the pre-processing stage, showing its individual elements and the geometric transformations aimed at simplifying the representation (e.g. the straightening transform).

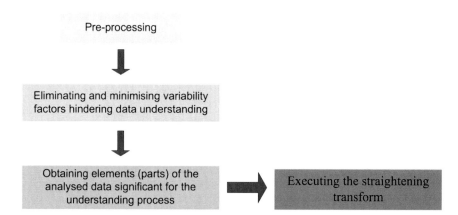

Fig. 2.21. Pre-processing in the process of understanding the analysed medical image

The presented approach to the analysis is a concept aimed at the semantic interpretation of medical images [87], [92], [97].

Pre-processing produces images cleaned of noise and reduced to such an extent, in the sense of eliminating superfluous details, that they become graphs of 'straightened' structures, which are then approximated using an adaptively selected broken line.

The above one-dimensional graphs mainly present information on the variability of the width and the profile of the edge of the organ in question. These graphs are used for understanding the contents of the analysed images under the assumption that in the medical problem under consideration this form of representation is sufficient to extract features of the image that are diagnostically significant. Obviously this may not be true for every type and form of images. However, this type of an analysis has found prior application to the automatic interpretation of a broad range of images from diagnostic examinations of organs in the abdominal cavity, the chest and the vascular (circulatory) system [87], [112]-[117], so there is every reason to also use it for pre-processing images of the spinal cord, bones of lower and upper extremities.

It is important to adapt linguistic methods of the cognitive analysis of medical images by generalising them for the structure of the nervous system and the skeleton of feet and hands, as the linguistic approach supports analysing morphological changes in the shape of anatomical structures of the spinal cord and bones of extremities. In addition, using grammars of sufficient generative strength also makes it possible to identify additional information (mainly semantic in nature) characterising the organ or structure examined.

Chapter 3
Cognitive Information Systems

Cognitive information systems were developed on the foundation of intelligent information systems whose purpose was not just the simple analysis of data consisting in recording, processing and interpreting it, but primarily an analysis by understanding and reasoning about the semantic contents of the processed data.

Every information system which analyses a selected type of data and information using certain features characteristic for them keeps, in its database, some knowledge – indispensable for executing the correct analysis – which forms the basis for generating the system's expectations as to all stages of the analysis it conducts. As a result of combining certain features found in the analysed type of data with the expectations – generated using this knowledge – about the existing semantic data/ information contained, cognitive resonance occurs as described previously.

Cognitive information systems utilise methods which define structural reasoning techniques serving to match patterns [92], [112]. A system which executes cognitive data analyses very often analyses not just information or numerical data, but also image data. In this last case, during the analysis process, the structure of the image being analysed is compared to the structure of the image serving as such a pattern. This comparison is conducted using strings of derivation rules which enable the pattern to be generated unambiguously. These rules, sometimes referred to as productions, are established as part of a specially derived grammar, which in turn defines a certain formal language, called an image language. An image (information) thus recognised is assigned to the class to which the pattern that represents it belongs.

Cognitive analysis utilised in cognitive information systems extremely frequently follows a syntactic approach which employs functional blocks for the semantic analysis and interpretation of the image [26], [30]. Pre-processing stages completed by coding the image, approximating the shapes of analysed objects, filtering and processing the image fed to the input of the system make it possible to obtain a new representation of the image presented as hierarchical structures of the semantic tree and subsequent steps of deriving this representation from the initial symbol of the grammar [30], [73].

During image data pre-processing, an intelligent cognitive recognition system must (in most cases) segment the image, identify picture primitives and also determine relations between them.

The classification proper (as well as the machine perception) consists in recognising whether the specific representation of the input image belongs to the class of images generated by the formal language defined by one of the grammars that can be introduced – a sequential, a tree or a graph grammar – which are used for recognition processes executed during the syntactic analysis performed by the system [30], [73], [92].

L. Ogiela and M.R. Ogiela: Advances in Cognitive Information Systems, COSMOS 17, pp. 51–60.
springerlink.com

The most recent studies of intelligent information systems indicate that only recognising the analysed image is no longer sufficient, because researchers increasingly frequently propose to employ these systems also for the automatic, computer understanding of the image. This applies in particular to image data containing layers of semantics. Such image data includes e.g. medical images. In order to enable reasoning for a selected class of patterns, artificial intelligence techniques are used. These techniques, apart from the simple recognition of the image identified for analysing, can also extract significant semantic information which facilitates its semantic interpretation, i.e. its full understanding.

This process applies only to cognitive information systems and is much more complex than just recognition, as the information flow in this case is clearly two-way [87], [92], [98], [105]. In this model, the stream of empirical data contained in the subsystem whose job it is to record and analyse the image interferes with the stream of expectations generated. A certain type of interference must occur between the stream of expectations generated by the specified hypothetical meaning of the image and the stream of data obtained by analysing the image currently under consideration. This interference means that some coincidences (of expectations and features found in the image) become more important, while others (both consistent and inconsistent) lose importance. This interference, which in consequence leads to cognitive resonance, confirms one of the possible hypotheses (in the case of an image whose contents can be understood), or justifies a statement that there is an irreparable inconsistency between the image currently perceived and all gnostic hypotheses which have an understandable interpretation. The second case means that the attempt to automatically understand the image has failed [96].

Thus the operation of cognitive information systems hinges on cognitive resonance which characterises only these systems and distinguishes them from other information systems. The use of such systems may be varied, as today's science offers wide-ranging opportunities for their application. The best applications for cognitive information systems are currently found in medicine, as more and more pathological entities are identified in disease processes of individual human organs and the capability to detect and recognise these entities keeps improving. Medical images are among the most varied forms of data and have extremely deep and significant meaning interpretations. Cognitive information systems can certainly also help in many other fields of science and everyday life if an attempt is made to add the process of understanding the analysed information/data to intelligent information systems in the fields of economics, marketing, management, logistics, defence and transport.

3.1 General Classification of Cognitive Information Systems

Cognitive analysis based on processes of learning about and understanding the studied phenomenon has become an opportunity for developing intelligent information systems almost in every field in which data is currently analysed. The general class of information systems using cognitive analysis for semantic interpretation and reasoning is referred to as cognitive categorisation systems, among which an important role is played by systems for understanding various types of information, as well as by data analysis and interpretation systems (Fig. 3.1).

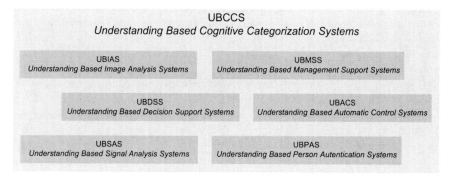

Fig. 3.1. Cognitive system division. Source: developed on the basis of [134], [136]

UBCCS cognitive categorisation systems include decision systems (UBDSS) developed to take decisions and reason based on them.

In the meaning of the classical decision theory, decision systems follow the described behaviour of an ideal, perfectly rational decision-maker (a person or a system) in circumstances which require one of a number of well-known options to be chosen [32], [38]. A cognitive system taking a decision is guided by the utility principle and considers all the options which somehow contribute to taking the right decision. The second principle which a cognitive decision system should follow in its operation is the principle of probability which rejects options (decisions) that are not realistic enough.

The classical decision theory makes several significant assumptions, of which the most important is that a cognitive decision system should have the full knowledge of all options available (to the system). It is nothing new that a great majority of decisions are taken in conditions of some risk, where it is uncertain what type of a possible decisions can be considered, and in particular what the consequences of taking a specific decision may be.

Cognitive decision systems are therefore developed to conduct analyses during which all the available options are considered and their suitability is compared in every regard. This can happen only if the system has the resources it needs, which include time and the computing capacity. With regard to these two inputs, it should be emphasised that it is them that can cause the decisions to be taken within the optimum time or can cause the system to conduct the decision analysis over a time exceeding common sense limits. In the later case, the decision taken will become useless, because the time needed to take the right decision counts in the decision-making process. A cognitive decision system analyses not only taking into the account the time available for completing the job, but also the ability to cognitively process all the available options including the consequences of adopting every one of them. A situation in which we have all the cognitive resources necessary to consider every option may apply in large, strong units (e.g. business ones).

Cognitive decision systems also follow the rule of the materiality of the adopted criterion. It is well known that not all decision-making criteria can be assigned the same weight in every case, as some of them are more material than others, even when this does may not seem desirable. In addition, some decision-making criteria may operate interactively, so they cannot be considered in total isolation [75].

Cognitive decision systems analyse and then execute decision-making processes using the classical decision theory, so they try to eliminate new approaches to the subject of decision-taking. These latter approaches are based on the current decision-making strategies associated with humans. The most important of them is the satisficing strategy which consists in reviewing the available options in a random order and selecting the first which suffices to satisfy the decision-maker [128], [129]. Human decision-making processes are not the best, as people taking decisions do not compare all the available options in every possible regard to select the best one. For a human, this is very often impracticable because of the shortage of time, knowledge, cognitive resources or computing capacity. According to Simon [129], such limitations are due to the concept of the bounded rationality of the human mind. The satisficing strategy is a very poor decision-making strategy, as it very frequently does not lead to expected (beneficial) results.

Another strategy followed by people when taking decisions, which is avoided by using cognitive decision systems, is the elimination by aspect strategy [139], which consists in determining a series of criteria and then eliminating those decisions that do not meet subsequent criteria. In this strategy, an option once rejected is no longer taken into account in subsequent considerations, even if some aspects of it are very appealing. Due to the above defects of the human decision-making process, cognitive UBDSS systems make use of only those human decision-making processes that can be considered optimum solutions. These processes must be quick in terms of the time taken to make the decision, and economical in terms of the complexity and difficulty of cognitive operations necessary in the decision-making process. Examples of such heuristics have been proposed in psychology, and they comprise the following rules [39]:

- The 'follow what is most important' rule – decisions are taken in accordance with the rule of following certain criteria, not all of which are of the same importance. The essence of this method is selecting the most important criterion, and then comparing individual options in pairs, every time rejecting the one whose value is lower or unknown in the light of the selected recommendation.
- The 'follow what has worked recently' rule – decisions are taken based on the criterion that was tried and worked in the most recent attempt of the same kind.

The presented decision-making criteria do not always allow the best decision to be taken, and what is more often entail quite a high risk and uncertainty. This is why the concept increasingly frequently considered to be the completely correct description of the decision making process is the one which takes into account the perspective theory [52]. The perspective theory is based on the framing effect, i.e. the impact of the mental representation of the decision-making problem on the content of decisions taken. The cognitive representation of a problem may change completely or to a great extent as a result of the words used to describe it or of taking into account the broader context in which the decision is taken [87], [92]. A person taking a decision compares to some extent by overlaying a kind of frame on the considered situation, which leads to a specific presentation of the problem and in consequence to taking a decision consistent with the framing process [77], [87], [92].

The operation of cognitive decision systems depends on many significant components of the human decision-making process. However, not all human aspects of decision-making processes can be applied in systems without restrictions, because not all the rules mentioned constitute optimum decision-making rules. Cognitive decision systems operate using significant elements of psychological theories transferred to the field of information systems. The cognitive decision-making process itself is based on a complex human psychological process made up of cognitive and motivational elements.

Cognitive categorisation systems, in addition to cognitive decision systems, also include UBMSS information management systems built for taking strategic decisions, for instance at enterprises. Information aspects play a significant role and are quite important for cognitive systems of this type. These aspects do not apply only to the information acquisition phase, but also to its correct processing and interpretation. UBMSS systems, just as other classes of cognitive categorisation systems, work by using cognitive resonance, but their additional benefit is that the cognitive system created within the information system is founded on the semantic network found in the semantic network model. This is why in this type of models the concepts connected with decision systems found in enterprises are kept permanently in the form of a hierarchical network structure made up of nodes and relations between them that link them. In particular nodes of the network, various concept representations are coded, and in addition features characterising these concepts are identified. These features are assigned to concept representations at the lowest possible level of the generality hierarchy. Introducing this type of a concept representation meets the proposal of cognitive operation economy in which the economy of the network structure can often be associated with the cost of its operation. The analysis, interpretation and reasoning processes executed by UBMSS systems facilitate the economical (from the financial point of view) use of data contained in the available knowledge base in which concept representations are coded. In a UBMSS system, the semantic network described contains, in addition to the concepts coded in the system and presented as network nodes, also the semantic relations that link particular nodes. In this case, the semantic relationship between the present concept representations is expressed as the sum of all links between their designations and features. Concepts that are closely linked are characterised by a value of their mutual links in the network, which value corresponds to a certain community containing all their features. These links, called network paths, are differentiated in the UBMSS system with regard to their weight, which means that the stronger the link between two concept representations, the greater the weight assigned to the path that links them. This allows an easier mutual activation in the data processing operation. The duality of semantic relations occurring between concepts has been taken into account in building UBMSS systems. These relations comprise:

- relations built on positive links, and
- relations built based on negative paths.

Semantic relations identified based on negative paths are significantly beneficial in the process of taking strategic decisions, because they demonstrate that the greater the semantic distance between concept representations, the faster the decision can be taken that the sentence which positively links the represented concepts is false.

The presented models of semantic networks found in UBMSS systems allow semantic relations occurring for defined concept representations to be identified.

These models can also be used to develop the next class of cognitive categorisation systems – UBPAS personal authentication systems - which include biometric identification systems as an instance. UBPAS systems are founded on two basic groups of memory examination methods, which include direct and indirect methods.

In direct methods, the system retrieves from its memory everything that it knows (so this is a knowledge retrieval method) or it recognises elements occurring in the previously applied or learned material within the set presented to it (this is a recognition method). In the recognition method, another possibility is assumed, namely that the system should classify the analysed data, should interpret it and reason about it. This approach to analysis processes in UBPAS systems means that the system executes a set of actions which cannot be correctly completed without previously gained (acquired) knowledge. So UBPAS systems operate using two sets of knowledge – overt and secret knowledge. This is because the system analyses not only the sets known to it before, but also elements that meet definition criteria not known to it previously, without disclosing the definition itself, but at the same time classifying and independently identifying elements that fulfil the definition (or do not fulfil it). This method of operation of UBPAS systems confirms to some extent the presence of secret knowledge based on two primary analysis criteria:

- the exclusivity criterion which means that the given analysis method allows only the knowledge that is used for executing a given task and necessary for it to be gained: in this case, the processes of data analysis and interpretation are limited only and solely to those of them which are to some extent independent of other elements not connected with completing the specific task.
- the sensitivity criterion which means that the type of analysis conducted is exhaustive enough to reveal the entire information held in the system knowledge base and of which the system is in a sense 'aware': the essence of this approach is that if the system holds only a part of the overt knowledge about the analysed

phenomenon, this knowledge, which has not been collected in the system (because the knowledge could not be accessed), should not be considered secret.

In the case of UBPAS systems, just like other classes of cognitive categorisation systems, we cannot accept the thesis of the complete separation of overt and secret knowledge. It should be emphasised that bases of overt knowledge definitely dominate in these systems, but we cannot completely rule out some contribution of secret knowledge when the knowledge bases are built by expert teams. This approach to public knowledge means that we can agree with Cleeremans [23], that knowledge is secret when it influences the processing, interpretation and analysis of information without being aware of its features. Regardless of the unawareness of its impact on analysis processes, secret knowledge is represented in the memory of UBPAS systems, so it can be retrieved from it and used in analysis, interpretation and reasoning processes.

The next class of UBCCS subsystems are UBSAS systems used for varied signal analysis, and UBACS systems which are automatic control systems. These two types of systems are only mentioned in this study, as their role and significance has been described in various scientific and research publications by other researchers [3], [28], [54], [58], [118], [146], [148], [153], [157], [159].

Apart from the UBDSS, UBMSS and UBPAS systems presented above, this book will propose UBIAS cognitive systems designed for analysing image-type data and used to analyse various types of image patterns, in particular medical images [87]-[89], [95]-[109]. UBIAS systems are the fruit of several years of the authors' research and their characteristics are the leading subject of chapter 5, where they will be discussed in greater depth.

All the types of cognitive categorisation systems presented before have wide-ranging applications, from economics, sociology and philosophy to technical and defence sciences, medicine or natural science.

All cognitive categorisation system types use methods of cognitive analysis in their operation to extend the capabilities of classical data analysis technologies in order to reason based on the semantics (meaning) of that data or the analysed information.

3.2   A Formal Perspective on Cognitive Categorisation Systems

Both the ongoing scientific research work and the rapid development of information systems allow a new class of systems – cognitive categorisation systems – to be introduced for analysing and interpreting data.

In this study, the authors propose the following definition of computerized cognitive categorisation.

'Cognitive categorisation systems' describe intelligent information systems designed for conducting in-depth data analyses based on the semantic contents of this data. Semantic analyses are conducted with algorithms for describing this data

based on the expert information possessed (for example in the form of knowledge
bases) and the processes of machine (computer) perception and understanding of
data performed with the use of e.g. mathematical linguistics.

In cognitive categorisation systems, the interpreted data, due to its semantics,
will also be described, analysed and used for reasoning, which may mean that the
analysed data will not only be correctly processed, but also learned and understood. The detailed description of the data analysis process taking place in cognitive categorisation systems is presented in Figure 3.2.

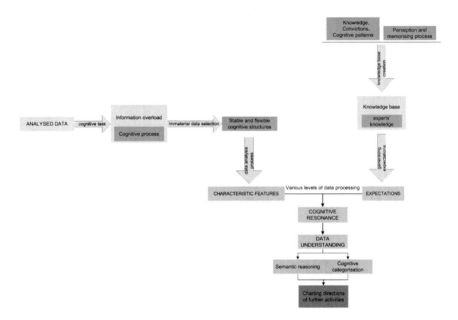

Fig. 3.2. The data analysis process in cognitive categorisation systems

Cognitive categorisation systems analyse data following human cognitive processes in which various information processing stages can be distinguished. In the
case of cognitive categorisation systems, information is not the only type of data
that can be subjected to cognitive categorisation processes which borrow the
methods of executing processes composed of various operations from classical information processing. Thanks to these processes, the course of the stimuli received
by the system (outside data) is optimised during the executed process. This process is equated to the assumption of the economical course of processes taking
place (necessary in the analysis). This is why significant cognitive tasks (for the
analysis process) are determined for the data undergoing analyses, which leads to
the correct description of the cognitive process. During the description of the correct cognitive process, an information overload can take place. This overload can
be eliminated only if a stage of selecting data that is immaterial (form the point of
view of the analysis conducted) is introduced. This selection, which is possible
because the system has a king of 'attention' directed at eliminating superfluous

information and because it repeats the same actions (operation) many times, leads to choosing increasingly stable and flexible cognitive structures.

The stage of information overload and superfluous data elimination can be equated with the stage of data pre-processing during which the information contained in the analysed data is reduced in such a way that significant information searched for is not lost, but at the same time this process finally produces a new data representation significant for the conducted analysis process.

Unlike the defined cognitive processes, cognitive structures are relatively permanent in the process of defining cognitive categorisation systems, and the system can use those structures many times in various conditions and situations.

Cognitive categorisation systems operate not just depending on the regularity of the cognitive processes executed to correctly classify the data type, but also based on a defined notion apparatus contained in the appropriate type of expert knowledge bases. This type of knowledge bases is created by expert teams using various knowledge elements, convictions and cognitive schemes, and are then utilised in cognitive data analysis and interpretation processes. The information collected by the system is used to confront (compare) the characteristic features it distinguished during the data analysis process with certain expectations which the system has generated based on expert knowledge. This comparison is possible at various layers of data processing, which means that the same data can be analysed with various intensities and care, i.e. at various layers of data processing. The extent to which our data will be analysed depends on the demands of the situation (e.g. the external one) or the cognitive problem formulated. The multi-layer processing and analysis of data is quite widespread, as it reflects one of the elementary features of the human mind. Regardless of the selected layer for data processing and analysis, the comparison of expectations and features characteristic for the analysed data leads to cognitive resonance, which causes the analysed data to be understood. Data understanding processes require understanding the semantic contents of the analysed data, so at this operating stage of the cognitive categorisation systems, semantic reasoning takes place (about the meaning conveyed by the analysed data).

The semantic reasoning stage parallels the cognitive categorisation process, as a result of which objects are gathered into certain groups whose common characteristic is the set of joint features assigned a verbal label serving as the linguistic representation of the specific group. The linguistic representation found in cognitive categorisation processes comes from a natural language, and when it is applied in system processes, it comes from the computer language.

Semantic reasoning and cognitive categorisation processes become the starting point for determining the directions of subsequent treatment of the analysed data (i.e. taking a decision for the future).

3.3 Properties of Cognitive Categorisation Systems

Referring to the definition, presented in the previous chapter, of cognitive categorisation systems used for data analysis, interpretation and reasoning based on its semantic contents, it should be emphasised that cognitive categorisation systems have various properties, of which the most important are as follows [87], [92]:

- A broad spectrum of analysed data - cognitive categorisation systems can analyse various data, because the cognitive methods based on cognitive resonance which are implemented in the system are universal;
- Wide-ranging opportunities to use cognitive categorisation systems in various scientific fields;
- The use of formalisms and tools from the area of informatics which support semantically-oriented cognitive reasoning with the use of structural artificial intelligence techniques;
- The use of the unity of cognition and action rule – the idea behind the operation of cognitive systems is to implement human cognitive processes which are used by categorisation systems, but at the same time combine these cognitive processes with the action which cognitive categorisation systems execute at their last operating stage, that is at the stage of determining the directions of subsequent activity;
- The ability to freely select the categorisation method depending on the chosen representation of the analysed data - the existing categorisation methods (listed below) can be applied in cognitive categorisation systems equally well:

  - comparing sets of defining features;
  - comparing sets of all features and establishing the proportion of defining features;
  - determining the semantic distance, i.e. comparing to the adopted standard or prototype;
  - comparing to the first specimen found;

- The use of knowledge bases made up of the appropriate expert knowledge collected;
- Accounting for the chronometries of systems which are based on the chronometry methods of the mind, i.e. determining the reaction time for analysing a given phenomenon in relation to the stage of transmitting information about the analysed data to the system, the stage at which the system processes this information and the stage at which the cognitive system programs the reaction and executes it;
- The ability to deploy cognitive categorisation systems taken from the scientific research domain in the practical world.

Cognitive categorisation systems are used to understand and analyse data. Their significant feature is that as a result of the broadly-understood data analysis conducted by the system, an in-depth interpretation of this data, and therefore also its cognitive categorisation, becomes possible. This type of data classification and description allows semantic information to be extracted, which makes it possible to reason at the stage of the data analysis conducted.

Chapter 4
Intelligent Cognitive Data Analysis Systems of the UBMSS Type as an Example of Cognitive Categorisation Systems

UBMSS systems, as cognitive data analysis systems, can be used not only to analyse the economic figures of a company, but can also complement the analysis of data for information in healthcare. Thus these systems become start supporting the financial and strategic analysis of medical establishments (hospitals, clinics, healthcare companies providing various medical services). A feature of UBMSS systems is that they conduct the financial analysis of a company using elements of cognitive data analysis.

This book discusses an example UBMSS system to illustrate how cognitive data analysis methods are used to interpret selected economic and financial ratios. UBMSS systems can be used for the cognitive analysis of economic ratios, particularly financial or macroeconomic ones. UBMSS systems can, for example, conduct economic and financial analyses using the following ratios [9]:

- liquidity ratios;
- cash-flow ratios;
- asset utilisation ratios;
- operating result ratios;
- equity structure ratios;
- capital adequacy structure ratios;
- market result ratios;
- rates of return;
- ratios for particular departments:
    - financial and accounting;
    - production;
    - marketing;
    - sales;
    - technical;
    - HR;
    - logistics.

Every ratio group mentioned above can undergo semantic analysis. This book presents an analysis of selected liquidity ratios and rates of return. Among liquidity ratios, the following are distinguished [9]:

- liquidity ratio;
- current ratio;

L. Ogiela and M.R. Ogiela: Advances in Cognitive Information Systems, COSMOS 17, pp. 61–73.
springerlink.com   © Springer-Verlag Berlin Heidelberg 2012

- cash ratio;
- quick ratio;
- sales to current assets;
- inventory to working capital;
- receivables turnover;
- average (receivables) collection period;
- overdue receivables collection period (days);
- days sales in receivables;
- investment in receivables;
- final balance;
- inventory turnover;
- inventory to sales;
- payables turnover;
- payables payment period (days);
- working capital (days);
- working capital productivity;
- cash reserve (days);
- current liabilities;
- overdue current liabilities to total current liabilities;
- working capital to debt;
- fixed assets to long-term liabilities;
- short-term to long-term debt;
- risk-weighted asset conversion ratio;
- COGS (*cost of goods sold*);
- EBIT (*earnings before deducting interest and taxes*);
- NPV (*net present value*);
- ACP (*average collection period*);
- TAT (*total asset turnover*);
- fixed asset productivity.

The following ratios are distinguished in the second group, i.e. rates of return, [9]:

- gross margin,
- profit margin,
- operating margin,
- return on net assets;
- return on total assets;
- ROA (*return on assets*);
- return on total gross assets;
- return on fixed assets;
- return on working capital;
- return on equity;
- ROE (*return on equity*);
- return on share capital;
- ROI (*return on investment*);

- ROIC (*return on invested capital*);
- ROS (*return on sales*);
- NPM (*net profit margin*);
- ROCE (*return on capital employed*);
- RONA (*return on net assets*);
- return on total assets;
- IRR (*internal rate of return*);
- WACC (*weighted average cost of capital*);
- EPS (*earning per share*);
- DPS (*dividends per share*);
- DPR (*dividend pay-out ratio*);
- PER (*price earning ratio*);
- DYR (*dividend-yield ratio*).

The above groups of economic and financial indicators is used by cognitive systems designed to conduct a semantic analysis of the standing of a given enterprise.

4.1 A UBMSS System for a Single-Factor Analysis of the NPV

The simplest type of UBMSS systems are those that analyse single ratios of a company. The system analysing the NPV (net present value), a selected liquidity ratio, presented in this chapter can be considered the most important of such systems. This type of single-factor analysis systems represents the simplest kind of UBMSS systems. An example of its operation is presented below.

The net present value, as a dynamic method, is used to assess the economic efficiency of a tangible investment. It is used when analysing cash flows discounted at the current discount rate. The NPV is the difference between discounted cash flows and the original investment and is calculated using the following formula [9]:

$$NPV = \sum_{t=1}^{n} \frac{CF_t}{(1+r)^t} - I_0$$

where:

CF_t – the value of cash flows within period t;
r – the discount rate;
I_0 – is the original investment;
t – subsequent periods of investment operation.

For the proposed UBMSS system conducting an analysis using the NPV, the following formal grammar was proposed:

$$G_{NPV} = (\Sigma_{N_{NPV}}, \Sigma_{T_{NPV}}, P_{NPV}, S_{NPV})$$

where:

$\Sigma_{N_{NPV}}$ – denotes the set of non-terminal symbols defined as follows:

$\Sigma_{N_{NPV}}$ ={RESULTS, W, ACCEPT, NOT_ACCEPT, A, B, C},

$\Sigma_{T_{NPV}}$ – denotes the set of terminal symbols defined as follows:

$\Sigma_{T_{NPV}}$ = {'a', 'b', 'c'}, and the individual elements of this set take the following values: a = {0%}, b ∈ (0%, 100%], c ∈ [-100%, 0%) (Figure 4.1).

Fig. 4.1. Terminal symbols for the G_{NPV} grammar

S_{NPV} – the start symbol of the grammar, $S_{NPV} \in \Sigma_{N_{NPV}}$, S_{NPV} = RESULTS,

P_{NPV} – a set of productions, defined as follows:

1. RESULTS→ ACCEPT | NOT_ACCEPT
2. ACCEPT→ W *//if (W = accept) final_decision := accept*
3. NOT_ACCEPT → W *//if (W = not accept) final_decision := not accept*
4. W → A *// w=results*
5. A → a *// decision:= accept*
6. B → b *// decision:= accept*
7. C → c *// decision:= not accept*

The analysis of economic ratios conducted by UBMSS systems supports using the semantic contents of the analysed data to determine the nature of that data, its impact on the current standing of the company and the extent of changes it causes to the company and its environment taking into account the information currently possessed. Such an analysis is possible due to semantic information contained in the analysed sets. This semantic information may relate to:

- the scale (value) of analysed economic ratios;
- the frequency of their changes;
- the way they change;
- the regularity with which ratio changes occur;
- the number of changes observed;
- the type of changes observed.

The example UBMSS system discussed here can conduct a cognitive analysis of selected financial and economic ratios, which will make it possible to take the best strategic decision for the selected (analysed) company. Figure 4.2 shows example results of the operation of the UBMSS system proposed for semantic analyses and interpretations stemming from understanding the selected analysed financial ratio

(NPV). Figure 4.2 shows a situation in which the NPV is positive (0,435), a value at which the investment should be made, so the situation is fully acceptable. This figure also shows an example of an analysis of the indicator equal to zero, in which case the UBMSS system also decides that the investment under consideration is acceptable, unlike when the NPV is negative. The last example shows a situation in which the UBMSS system does not accept the investment.

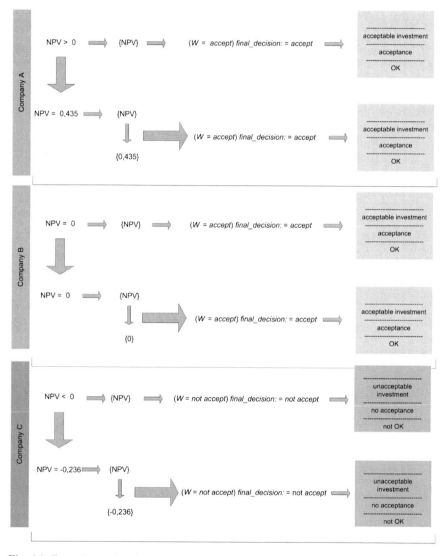

Fig. 4.2. Example results of the NPV analysis by a UBMSS system

The presented UBMSS system is an example of a simple cognitive system, but it still shows what the semantic interpretation of the value of the analysed economic indicator, a liquidity ratio, looks like. Another example of such systems is the analysis process for a selected rate of return.

4.2 An Example UBMSS System for a Single-Factor Analysis of the IRR

Another type of UBMSS systems analyses single rates of return of a company. The system analysing the IRR (internal rate of return), presented in this chapter, can be considered the most important of such systems.

The IRR is a complex method of assessing investment projects by analysing the interest rate and the (changing) time value of money, the risk of completing the specific investment as well as inflation.

The IRR is calculated using the following formula [9]:

$$IRR = d_1 + \frac{PNPV\,(d_2 - d_1)}{PNPV + |NNPV|}$$

where:

d_1 – discount rate for a positive NPV close to zero;
d_2 – discount rate for a negative NPV close to zero;
$PNPV$ – positive NPV close to zero;
$NNPV$ – negative NPV close to zero.

For the proposed UBMSS system conducting an analysis using the IRR, the following formal grammar was proposed:

$$G_{IRR} = (\Sigma_{N_{IRR}}, \Sigma_{T_{IRR}}, P_{IRR}, S_{IRR})$$

where:

$\Sigma_{N_{IRR}}$ – denotes the set of non-terminal symbols defined as follows:

$\Sigma_{N_{IRR}}$ = {RESULTS, W, STRONG_ACCEPT, ACCEPT, WEAK_ACCEPT, NOT_ACCEPT, A, B, C, D, E},

$\Sigma_{T_{IRR}}$ – denotes the set of terminal symbols defined as follows:

$\Sigma_{T_{IRR}}$ = {'a', 'b', 'c', 'd', 'e'}, and the individual elements of this set take the following values: a = {0%}, b ∈ (0%, 15%], c ∈ (15%, 45%), d ∈ [45%, 100%], e ∈ [-100%, 0%) (Fig. 4.3).

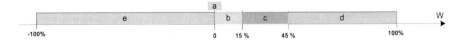

Fig. 4.3. Terminal symbols for the G_{IRR} grammar

S_{IRR} – the start symbol of the grammar, $S_{IRR} \in \Sigma_{N_{IRR}}$, S_{IRR} = RESULTS,

P_{IRR} – a set of productions, defined as follows:

1. RESULTS→ WEAK_ACCEPT | ACCEPT | STRONG_ACCEPT | NOT_ACCEPT
2. WEAK_ACCEPT→ W //if (W = weak accept) final_decision:= weak accept
3. ACCEPT→ W //if (W = accept) final_decision := accept
4. STRONG_ACCEPT → W //if (W = strong accept) final_decision := strong accept
5. NOT_ACCEPT → W //if (W = not accept) final_decision := not accept
6. W → A | B | C | D | E // W= results
7. A → a // decision:= weak
8. B → b // decision:= weak
9. C → c // decision:= accept
10. D → d // decision:= strong
11. E → e // decision:= not accept

The example UBMSS system discussed here can cognitively analyse the selected rate of return, making it possible to take the best strategic decision for the company subject to the analysis. Figure 4.4 shows example results of the operation of the UBMSS system proposed for semantic analyses and interpretations stemming from understanding the significance of the financial ratio analysed (IRR). Figure 4.4 shows a situation in which the IRR is positive at 0,09, 0,42 and 0,93, values at which the investment should be undertaken, although the level of acceptance is dramatically different in these three cases. In the first case, the investment is only just acceptable, in the second it is moderately acceptable, whereas in the third it is highly acceptable. In addition, an example analysis of the indicator equal to zero is presented. In this case, the UBMSS also finds the investment only just acceptable. What is more, a case in which the IRR is negative has also been analysed. In this case the investment is unacceptable.

The presented UBMSS system is an example of a simple cognitive system, but it still shows what the semantic interpretation of the value of the selected economic indicator, namely a rate of return, looks like.

The above two examples of the semantic analysis of financial and economic indicators show the semantic type of a ratio analysis. Every presented example of an UBMSS system may be extended to include other financial ratios significant for the economic and financial analysis of a given organisation.

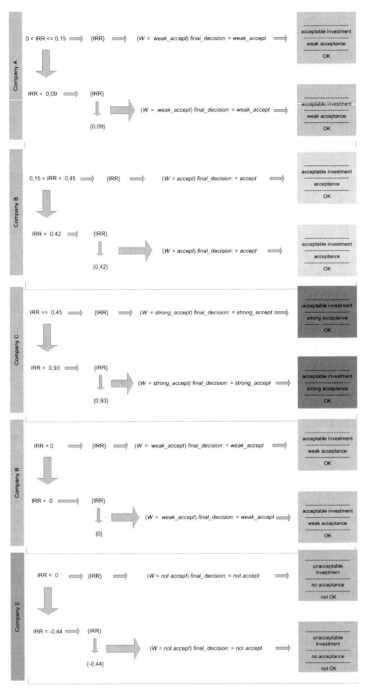

Fig. 4.4. Example results of an IRR analysis by a UBMSS system

4.3 An Example UBMSS System for a Dual Factor Analysis of the IRR and the Discount Rate *r*

Below, the reader will find a description of an UBMSS system interpreting two selected economic indicators, namely:

- *r* – discount rate (symbol: W1),
- IRR – internal rate of return (symbol: W2).

The discount rate is the rate at which future cash would be exchanged for cash available now. The discount rate stems from the changing time value of money. It shows to what extent future capital is different from the effective value of current capital. The discount rate is defined as the percentage rate of the discount amount to the future value of capital.

For the proposed UBMSS system conducting a dual-factor analysis of the indicators, the following formal grammar has been defined:

$$G_{IRRr} = (\Sigma_{N_{IRRr}}, \Sigma_{T_{IRRr}}, P_{IRRr}, S_{IRRr})$$

where:

$\Sigma_{N_{IRRr}}$ – denotes the set of non-terminal symbols defined as follows:

$\Sigma_{N_{IRRr}}$ ={RESULTS, W1, W2, WEAK_ACCEPT, ACCEPT, STRONG_ACCEPT, NOT_ACCEPT, A, B, C, D, E},

$\Sigma_{T_{IRRr}}$ – denotes the set of terminal symbols defined as follows:

$\Sigma_{T_{IRRr}}$ ={'a', 'b', 'c', 'd', 'e'}, and the individual elements of this set take the following values: a = {0%}, b ∈ (0%, 15%], c ∈ (15%, 45%), d ∈ [45%, 100%], e ∈ [-100%, 0%) (Fig. 4.5).

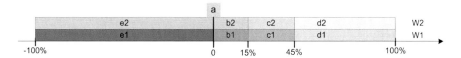

Fig. 4.5. Terminal symbols for the G_{IRRr} grammar

S_{IRRr} – the start symbol of the grammar, $S_{IRRr} \in \Sigma_{N_{IRRr}}$, S_{IRRr} = RESULTS,

P_{IRRr} – a set of productions, defined as follows:

1. RESULTS→ WEAK_ACCEPT | ACCEPT | STRONG_ACCEPT | NOT_ACCEPT
2. WEAK_ACCEPT→ W1 W2 //if (w1 & w2 = weak accept) final_decision:= weak accept
3. ACCEPT→ W1 W2 //if (w1 & w2 = accept) final_decision := accept

4. STRONG_ACCEPT → W1 W2 //if (w1 & w2 = strong accept) final_decision := strong accept
5. NOT_ACCEPT → W1 W2 //if (w1 & w2 =not accept) final_decision := not accept
6. W1 → A I B I C I D I E // w1=decision
7. W2 → A I B I C I D I E // w2=decision
8. A → a // decision:= weak
9. B → b // decision:= weak
10.C → c // decision:= accept
11.D → d // decision:= strong
12.E → e // decision:= not accept

An example UBMSS system analysing the selected economic indicators is shown in Figure 4.6.

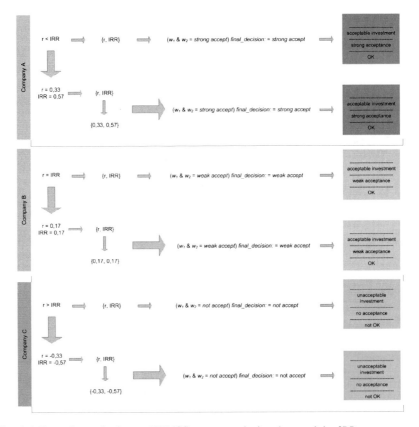

Fig. 4.6. Example results from a UBMSS system analysing the *r* and the *IRR*

Figure 4.6 shows a situation where the *r* and the *IRR* are different. In the first case, in which *r* is 0,33 and the *IRR* is 0,57, indicator values show that the investment should be undertaken, so this is a fully acceptable situation.

In the second case, r is equal to *IRR* at 0,17. In this case, the UBMSS also finds the investment acceptable.

The third case is where $r >$ IRR and both indicators take negative values. This case represents an unacceptable investment.

The results presented show the decision taken by the UBMSS after analysing not only the rates, but also the relationship between them. This kind of analysis can be undertaken if the significance of a change in one rate on the change in the second rate is determined.

Another example of a UBMSS system is presented in chapter 4.4 below, which discusses such a system analysing three selected economic and financial ratios.

4.4 An Example UBMSS System for a Multi-factor Analysis of the Economic and Financial Ratios

This chapter presents a UBMSS system interpreting three selected economic indicators, namely:

- NPV – net present value (symbol: W1);
- r – discount rate (symbol: W2);
- IRR – internal rate of return (symbol: W3).

For the proposed UBMSS system, the following formal grammar has been defined:

$$G_w = (\Sigma_{N_w}, \Sigma_{T_w}, P_w, S_w)$$

where:

Σ_{N_w} – denotes the set of non-terminal symbols defined as follows:

Σ_{N_w} ={RESULTS, W1, W2, W3, WEAK_ACCEPT, ACCEPT, STRONG_ACCEPT, NOT_ACCEPT, A, B, C, D, E},

Σ_{T_w} – denotes the set of terminal symbols defined as follows:

Σ_{T_w} ={'a', 'b', 'c', 'd', 'e'}, and the individual elements of this set take the following values: a = {0%}, b $\in$ (0%, 15%], c $\in$ (15%, 45%), d $\in$ [45%, 100%], e $\in$ [-100%, 0%) (Fig. 4.7).

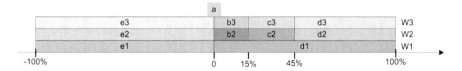

Fig. 4.7. Terminal symbols for the G_w grammar

S_w – the start symbol of the grammar, $S_w \in \Sigma_{N_w}$, $S_w = $ RESULTS,

P_w – a set of productions, defined as follows:

1. RESULTS→ WEAK_ACCEPT I ACCEPT I STRONG_ACCEPT I NOT_ACCEPT
2. WEAK_ACCEPT→ W1 W2 W3 //if (w1 & w2 & w3 = weak accept) final_decision:= weak accept
3. ACCEPT→ W1 W2 W3 //if (w1 & w2 & w3 = accept) final_decision := accept
4. STRONG_ACCEPT → W1 W2 W3 //if (w1 & w2 & w3 = strong accept) final_decision := strong accept
5. NOT_ACCEPT → W1 W2 W3 //if (w1 & w2 & w3 =not accept) final_decision := not accept
6. W1 → A I B I C I D I E // w1=decision
7. W2 → A I B I C I D I E // w2=decision
8. W3 → A I B I C I D I E // w3=decision
9. A → a // decision:= weak
10. B → b // decision:= weak
11. C → c // decision:= accept
12. D → d // decision:= strong
13. E → e // decision:= not accept

An example UBMSS system analysing the selected economic indicators is shown in Figure 4.8.

This example shows three fundamentally different situations. The first is an attempt to take a decision when the NPV is positive and the IRR is greater than the r. In this situation, the system finds the investment acceptable.

In the second case, the NPV is zero while the r is equal to the IRR. In this situation the system decides that the investment is only just acceptable because the r and the IRR are low.

The third case illustrates a situation in which the NPV is negative and the IRR is lower than the r. The UBMSS finds the investment unacceptable because all indicators are negative.

The semantic analysis conducted by a UBMSS system shows how important it is not just to analyse the indicators, but primarily to determine the dependencies between them. This is because an analysis should not be conducted using just the values of ratios. This could mean that the activities would produce wrong results since the analysis excludes the important impact other information has on the specific (analysed) ratio. The analysed ratio can be changed by many different factors, including the changes of other indicators influencing the value of the analysed ones.

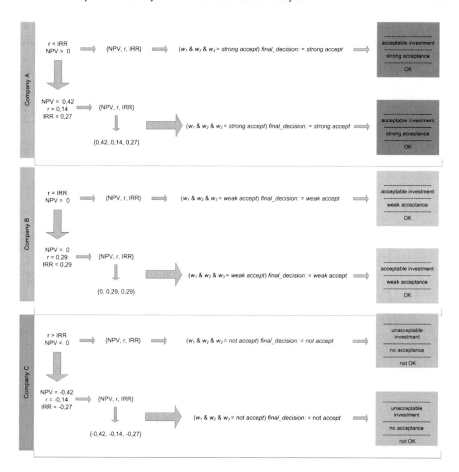

Fig. 4.8. Example results of an *NPV*, *r* and *IRR* analysis by a UBMSS system

All cases of the analysed economic and financial ratios demonstrate how immensely important it is to determine their impact on and significance for the analysed investment or the financial standing of the company when the system is taking the best decision. The presented examples of the cognitive analysis of economic data and its meaning-based interpretation allow the right decision to be taken as to the acceptability of an investment forming the subject of complex decision-making analysis processes.

Chapter 5
UBIAS – Intelligent Cognitive Systems for Visual Data Analysis

UBIAS image analysis systems are used to analyse data in the form of images and have been developed for a number of years as part of work to apply cognitive data analysis systems in practical, automated meaning interpretation tasks.

The characteristic feature of such systems is their universality and high precision when they execute interpretation and decision-making processes. Image analysis systems were initially developed as one class of cognitive data categorisation systems and, being the first, became the starting point for writing the complete classification of cognitive data analysis systems.

Image analysis systems, because they analyse completely different data sets, meaning not just different types of images, but also disparate features of them, make it possible to introduce newer and newer definitions and thus to modify (improve) solutions deployed earlier.

UBIAS systems analysing medical images are considered to be fundamental and were the original reason for proposing this class of cognitive data categorisation systems which have been very successfully developed for many years in the following publications [82]-[92], [97], [114], [136].

The essence of the operation of UBIAS systems is the correctly conducted decision-making process based on cognitive resonance which looks as follows for UBIAS systems (Figure 5.1).

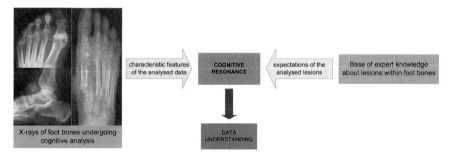

Fig. 5.1. Cognitive resonance in UBIAS systems

The medical images selected for analysing, showing lesions within various human organs, undergo a semantic analysis aimed at isolating anomalies significantly impacting the patient's life and health from the analysed sets. Based on the images analysed, sets of features characteristic for the discussed images are defined. These sets are compared to expectations about the analysed data formulated

L. Ogiela and M.R. Ogiela: Advances in Cognitive Information Systems, COSMOS 17, pp. 75–83.
springerlink.com   © Springer-Verlag Berlin Heidelberg 2012

on the basis of the expert knowledge base collected in the system. Figure 5.1 shows an example of cognitive resonance occurring in the analysis of bone images of a foot in which lesions are detected. When the characteristic features of the analysed foot bone X-rays are compared to expectations concerning lesions found in the X-ray images analysed, a resonance occurs, as a result of which consistencies of pairs of features and expectations are determined, or inconsistencies between the characteristic features and expectations are identified. Only the first case allows further cognitive activity leading to understanding the analysed data to be carried out at this analysis stage.

UBIAS systems illustrate the analysis of image data, and in particular various types of medical images. Such systems are designed to semantically analyse data in the form of various lesions found in different parts of different organs. Below, an example UBIAS system used for analysing and interpreting image data obtained from X-rays of foot bones is presented.

This system was developed to analyse and interpret image data associated lesions occurring within the foot skeleton.

The following lesions were distinguished in the lesion analysis:

- toe fractures;
- foot bone fractures;
- degenerative lesions;
- foot deformations;
- arthritis;
- tuberculosis of bones.

A cognitive data analysis system analysing lesions occurring within foot bones shown in X-ray images is an example of a cognitive system applied to the semantic analysis of image data.

The UBIAS system carries the analysis out by using mathematical linguistic algorithms based on graph formalisms proposed in publications [92], [112].

The key aspect in introducing the right definition of the formal grammar is to adopt names of bones found within the foot, which include:

- talus (tal),
- calcaneus (calc),
- cuboid – os cuboideum (cubo),
- navicular – os naviculare (navi),
- lateral cuneiform – os cuneiforme laterale (cunei_l),
- medial cuneiform – os cuneiforme mediale (cunei_m),
- middle (intermediate) cuneiform – os cuneiforme intermedium (cunei_i),
- metatarsal – os metatarsale (m),
- hallux (h),
- proximal phalanx – phalanx proximalis (phalanx_prox),
- middle phalanx – phalanx media (phalanx_med),
- distal phalanx – phalanx distalis (phalanx_dist),

The healthy structure of foot bones is shown in Figure 5.2 (a, b).

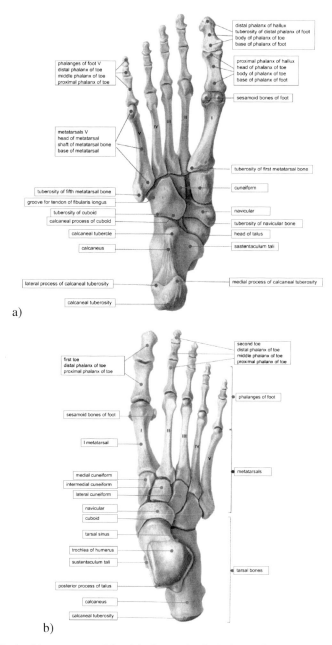

Fig. 5.2. The healthy structure of the right foot – a) sole (palmar) side, b) dorsal side

Figure 5.2 presents all the foot bones, divided into metatarsus, tarsus and phalanx bones, which will be analysed by the proposed UBIAS system.

To provide the right insight into the proposed way of analysing foot bone lesions in X-ray images, an X-ray of a foot free of any lesions within it, which represents the pattern defined in the UBIAS system, is shown below (Fig. 5.3.).

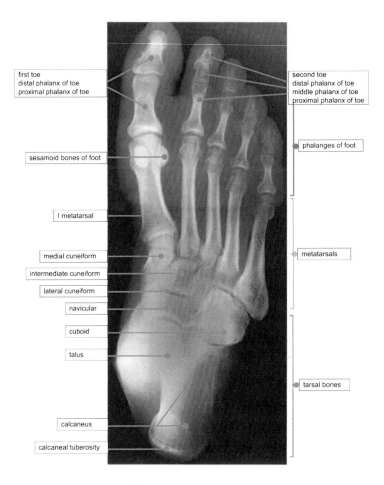

Fig. 5.3. Correct foot structure – X-ray

The example UBIAS system discussed in this chapter is used for the image analysis of foot bone lesions in the dorsopalmar projection of the foot, including the division of bones into tarsus, metatarsus and phalanx ones.

The grammatical graph formalism for the semantic analysis has been defined as the G_{foot} grammar taking the following form:

$$G_{foot} = (N_f, T_f, \Gamma_f, S, P)$$

where:

N_f – denotes the set of non-terminal symbols defined as follows:

N_f = ={ST, CALCANEUS, TALUS, CUBOIDEUM, NAVICULARE,
CUNEI_LATERALE, CUNEI_MEDIALE, CUNEI_INTERMEDIUM,
METATARS1, METATARS2, METATARS3, METATARS4, METATARS5,
HALUX1, HALUX2, PHALANX2_PROX, PHALANX2_MED,
PHALANX2_DIST, PHALANX3_PROX, PHALANX3_MED,
PHALANX3_DIST, PHALANX4_PROX, PHALANX4_MED,
PHALANX4_DIST, PHALANX5_PROX, PHALANX5_MED,
PHALANX5_DIST}

T_f – denotes the set of terminal symbols defined as follows:

T_f = {calc, tal, cubo, navi, cunei_l, cunei_m, cunei_i, m1, m2, m3, m4, m5, h1,
h2, p2_p, p2_m, p2_d, p3_p, p3_m, p3_d, p4_p, p4_m, p4_d, p5_p, p5_m, p5_d}

Γ_f – {p, q, r, s, t, u, v, w, x, y, z} – presented in Figure 5.4.

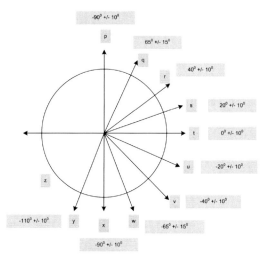

Fig. 5.4. Elements of the Γ_f set

S – the start symbol of the grammar,
P – the finite set of productions defined in Figure 5.5.

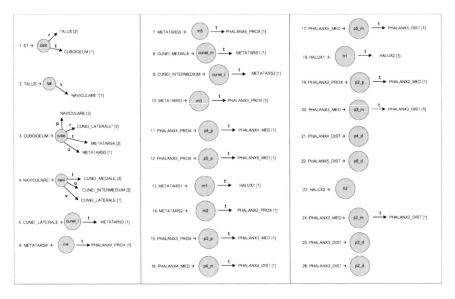

Fig. 5.5. Set of productions *P*

Figure 5.6, in turn, shows the graph of relations between individual tarsus, metatarsus and phalanx bones.

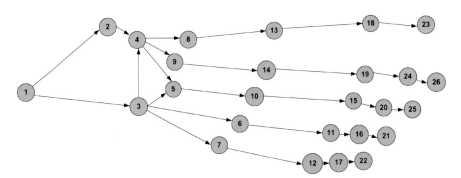

Fig. 5.6. The graph of relations between individual foot bones

Figure 5.7 shows the graph of relations between individual foot bones including the angles of slopes between them.

This definition method allows the UBIAS system to start analysing image data. Selected results of its operation are illustrated by Figures 5.8-5.12, which show the universality of the analysis using various selected examples of automatic image data interpretation and its semantic interpretation.

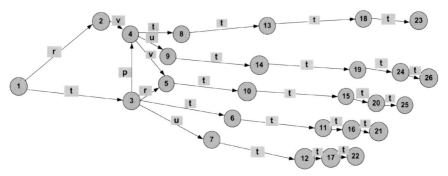

Fig. 5.7. A graph of relations between individual foot bones including the angles of slopes between them

This should be compared to the following medical images used for the cognitive data interpretation with the application of graph formalisms to analyse images showing various foot bone lesions.

Figure 5.8 shows an example of the automated analysis of an image showing a foot deformation caused by uric acid diathesis, colloquially called gout.

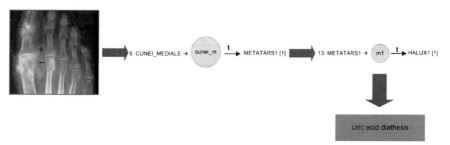

Fig. 5.8. Image data analysis by UBIAS systems to understand data showing uric acid diathesis

Figure 5.9 shows an example of the automatic analysis of an image depicting a fracture of *os naviculare*.

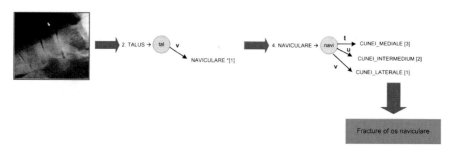

Fig. 5.9. Image data analysis by UBIAS systems to understand data showing a fracture of os naviculare

Figure 5.10 illustrates the method of UBIAS system operation when this system analyses image data to automatically understand data showing a foot deformation.

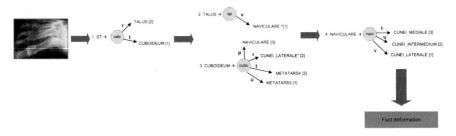

Fig. 5.10. Image data analysis by UBIAS systems to understand data showing a foot skeleton deformation

Figure 5.11 is an example attempt to automatically analyse image data illustrating a fracture of the neck of the talus of the foot.

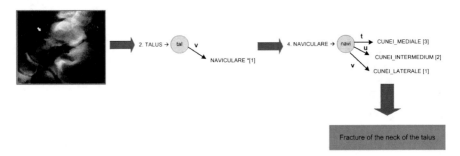

Fig. 5.11. Image data analysis by UBIAS systems to understand data showing a fracture of the neck of the talus

Figure 5.12 is an image data analysis carried out by an UBIAS system on an example showing a foot deformation caused by diabetes.

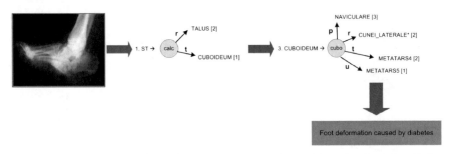

Fig. 5.12. Image data analysis by UBIAS systems to understand data showing a foot skeleton deformation caused by diabetes.

All the above examples of automatic image data analysis demonstrate the essence of UBIAS cognitive system operation, namely the correct understanding of the analysed lesion using series of productions defined in the system and the semantics of the analysed images.

Systems for the automatic analysis and interpretation of image data carry out a process of understanding the analysed data using the semantic information contained in them and defined in production sets of the formal grammar implemented in the system. The introduction of semantic records in sets of productions makes it possible to semantically analyse image data, whose interpretation and significance is of the greatest importance in cases of progressive lesions.

Chapter 6
E-UBIAS – Cognitive Systems for Image and Biometric Data Analysis

UBIAS systems which conduct semantic analyses are also capable of biometric identification using the following characteristics:

- individual;
- physical;
- behavioural.

The main individual characteristics are determined for every image analysed. However, if they concern a specific case, then they have to be identified in the base in which they are defined as identifiers of that case. For instance, if what is analysed is an image of the hand of someone with one finger much shorter in the right hand, then the definition of the pattern for that person should contain a correctly defined pattern of that person's hand covering finger lengths, including the unusual length of one of them.

However, if the item defined is the shape of the foot of a given individual for whom the length of the first toe of the right foot is unusual, then this aspect has to be included when defining the foot pattern. Otherwise it will turn out that the traditional patterns of the hand or the foot defined in the cognitive data analysis system do not facilitate the correct semantic analysis due to the disparities between the pattern and the specific individual.

A biometric analysis can also be carried out using the physical features of a given individual. Basic physical traits that are used for biometric analyses include:

- fingerprints;
- iris;
- retina;
- face shape;
- mouth shape;
- ear shape;
- hand shape;
- palm lines;
- geometry of the vascular system of a given body organ;
- body temperature;
- DNA;
- smell;
- timbre.

Biometric analyses are also carried out using behavioural traits, which now mainly include:

- voice analysis of the person (analysing their speech, singing or shouting);
- handwriting analysis;

L. Ogiela and M.R. Ogiela: Advances in Cognitive Information Systems, COSMOS 17, pp. 85–97.
springerlink.com

- gait analysis;
- typing analysis (e.g. on a computer keyboard);
- analysis of human brain reactions.

The authors believe that the biometric analysis of the hand deserves particular attention, as examples show how often our identity is analysed based on the shape and characteristic features of our hands. We most often encounter this type of analysis during security checks at airports of many countries where we are requested to place our fingers (or the whole hand) on a scanner which takes a photograph of our hand and records its characteristic features in the database of the personal identification system. In addition, our photograph, also taken during routine passport checks at airports, is attached to this recorded image of the hand or a single finger.

The human hand is characterised by many features of an individual nature, which include:

- hand shape;
- finger spans;
- sizes of areas between individual hand bones;
- finger lengths;
- relative locations of finger bones;
- metacarpus bone shape;
- metacarpus bone lengths;
- relative locations of metacarpus bones;
- wrist bone shape;
- wrist bone sizes;
- relative wrist bone locations;
- relative hand bone locations (finger, metacarpus, wrist bones);
- thicknesses of bones.

The image of a healthy hand is shown in Figures 6.1 and 6.2 which present the bones of the right hand – of the fingers, the metacarpus and the wrist.

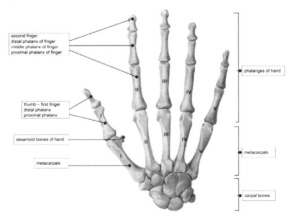

Fig. 6.1. The healthy hand structure – view of right hand bones from behind

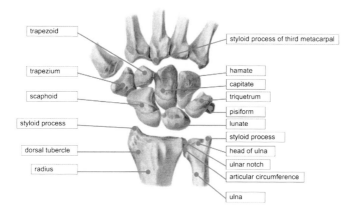

Fig. 6.2. The healthy structure of right wrist bones – view from behind

The structure of wrist bones shown in Figures 6.1 and 6.2 looks as follows on X-rays (Figures 6.3 and 6.4).

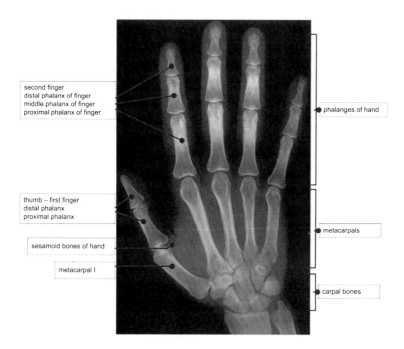

Fig. 6.3. An X-ray of a healthy hand structure – view of right hand bones from behind

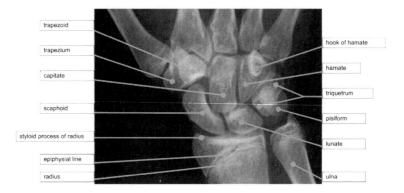

Fig. 6.4. An X-ray of the healthy structure of right wrist bones – view from behind

The hand bone structure of either one selected hand or both hands can form the basis of a biometric analysis as differences in the hand structure can be present in either both hands or just one. This analysis consists in measurements taken with a special reader which, during the procedure, records a 3D image of the hand, storing its characteristic traits. These records are kept in the database as patterns assigned to individual people. Every such pattern is unique and unanimously identifies the person to which it is assigned. If there is no pattern in the system for the person currently undergoing the biometric analysis, this may lead to either no pattern being assigned to the person, or to his/her wrong identification. The wrong identification may also occur if the biometric pattern stored in the database is not updated, because as the time passes and the person ages, his/her hand also ages and the following changes may occur in the hand:

- hand shape changes;
- changes of its span;
- changes in the size of areas between bones.

Modern biometric systems account for changes in the hand geometry caused by the time factor (the passage of time).

What seems particularly interesting is research combining the above traditional biometric analysis technologies with cognitive analysis techniques. This is because techniques consisting in the semantic analysis of the hand shape based on the analysis of lesions (or the lack thereof) found within the hand can be supplemented with a biometric analysis.

Individual records contained in cognitive systems may contain (assign) information on hand biometrics, i.e. its shape, the length of particular hand bones, bone thicknesses, the relative location of bones and the hand span. It should be noted that the cognitive analysis of hand bone images is based on selecting the right linguistic formalism for a graph description of hand bones. This description is written based on the healthy location of hand bones for:

- the centres of gravity of individual hand bones;
- the beginning and the end of every hand bone.

In both cases, hand geometry is unanimously identified by the length of individual hand bones. For this notation, to identify the specific person, the records containing model descriptions can be complemented with individual biometric characteristics and the physical features, and in the case of very complex semantic processes, behavioural traits can also be added. Complementing UBIAS systems with biometric analysis processes has given birth to a new class of semantic analysis systems called E-UBIAS (*Extended Understanding Based Image Analysis Systems*). This class of systems is designed for executing identification analyses, semantic/biometric analyses and recognition analyses.

E-UBIAS systems carry this semantic analysis out as described below. The identification system records the hand image for which the characteristic traits of the hand are saved in an individual record. A diagram of characteristic traits assigned to a hand is illustrated in Figure 6.5, showing the introduced definitions of particular measures and values assigned to hand geometry.

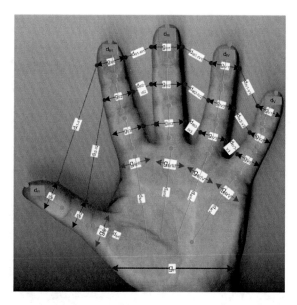

Fig. 6.5. Characteristic features of hand geometry

For the correctly described geometry of the hand, a record is defined in which data is recorded in the form of a set:

$$B_{dl} = \{ g_{ij}, d_{ij}, d_{ij\text{-}ij}, g_{śri}, d_{śri}, g_n \}$$

where:

g_{ij} – denotes the thickness of the bones of the ith finger and the jth phalanx for i = {I, II, III, IV, V}, j = {1, 2, 3}

d_{ij} – denotes the length of the bones of the ith finger and the jth phalanx for i = {I, II, III, IV, V}, j = {1, 2, 3}

$d_{ij\text{-}ij}$ – is the size of areas between individual hand bones,

$g_{\acute{s}ri}$ – is the thickness of the ith metacarpus bone, for $i = \{I, II, III, IV, V\}$,

$d_{\acute{s}ri}$ – is the length of the ith metacarpus bone,

g_n – is the size of wrist bones.

Introducing a formalism for the biometric notation of hand geometry makes it possible to precisely identify the person to whom a given vector of data has been assigned. At the same time, the biometric traits that characterise the specific person can be precisely established. The biometric feature vector can be complemented with new data presenting:

- the shape of the fingerprints of this hand;
- the shape of the lines of the palm;
- the shape of the complete handprint.

This situation is shown in Figure 6.6 in which the shapes of fingerprints of particular fingers are shown in blue, and the shape of palm lines in green. All the areas marked illustrate the third case, i.e. the shape of the entire handprint.

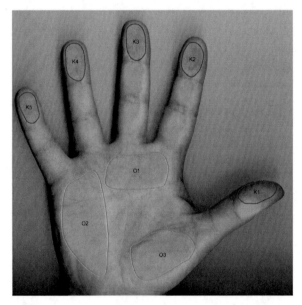

Fig. 6.6. Characteristic features of hand geometry

For an example of biometric features thus illustrated, the following formalism for the biometric recording of physical features of hand bones was defined. It has the following form:

$$L_{dl} = \{K_i, O_j\}, i = \{1,...,5\}, j = \{1,...,3\}$$

where:

K_i – denotes the print of the ith finger of the hand (from one to five),

O_j – the shape of one of the three biometric prints of the palm.

Introducing a formalism defining the B_{dl} set and additionally also the L_{dl} set serves not just to correctly verify and identify the given person as shown in Figure 6.7, but also to understand, as such, this image data.

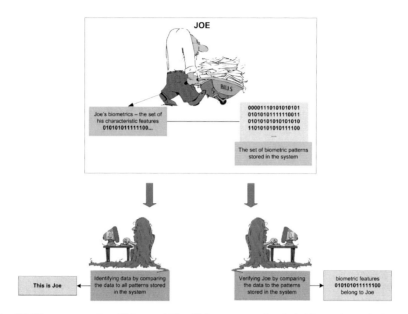

Fig. 6.7. The process of verifying and identifying a person by their biometric analysis

The essence of the biometric analysis is the correct definition of a set of biometric traits based on which a system database is built, which contains all the biometric patterns defined in the system. The verification process consists in comparing selected data with the patterns stored in the system. On this basis, the system determines whether the data corresponds to the pattern stored in the system or is inconsistent with it. So as a result of the system operation, it is possible to determine whether the analysed biometric data belongs to a given person or not. In the identification process, it is possible to identify who the person whose biometrics has been analysed is.

The combination of the biometric and semantic analyses makes it possible to cognitively/semantically verify and identify persons. These aspects of cognitive analysis complement identification processes by adding the functionality of semantically interpreting the results obtained. This is why, in addition to the presented biometric analysis of the hand, the hand bone image is semantically analysed in parallel.

The semantic analysis of X-rays of hand bones is based on mechanisms of linguistically perceiving and understanding data defined in the form of a formal grammar. The key in introducing the right definition of the formal grammar is to adopt names of bones found within the palm, which include the following:

- radius (rs),
- ulna (ul),
- scaphoid – os scaphoideum (sc),
- lunate – os lunatum (lt),
- triquetrum – os triquetrum (tq),
- pisiform – os pisiforme (pf),
- trapezium – os trapezium (tz),
- trapezoid – os trapezoideum (tm),
- capitate – os capitatum (c),
- hamate – os hamatum (h),
- metacarpals – ossa metacarpale (m),
- sesamoid bones of hand – ossa sesamoidea (ses),
- proximal phalanx – phalanx proximalis (pip),
- middle phalanx – phalanx media (pm),
- distal phalanx – phalanx distalis (pd).

The grammatical graph formalism for the semantic analysis has been defined as the G_{hand} grammar taking the following form:

$$G_{hand} = (N_h, T_h, \Gamma_h, S_h, P_h)$$

where:

N_h = {ST, RADIUS, ULNA, OS SCAPHOIDEUM, OS LUNATUM, OS TRIQUETRUM, OS PISIFORME, OS TRAPEZIUM, OS TRAPEZOIDEUM, OS CAPITATUM, OS HAMATUM, M1, M2, M3, M4, M5, SES1, SES2, PIP1, PIP2, PIP3, PIP4, PIP5, PM2, PM3, PM4, PM5, PD1, PD2, PD3, PD4, PD5} denotes the non-terminal symbol set,

T_h – denotes the set of terminal symbols defined as follows:

T_h = {rs, ul, sc, lt, tq, pf, tz, tm, c, h, m1, m2, m3, m4, m5, ses1, ses2, pip1, pip2, pip3, pip4, pip5, pm2, pm3, pm4, pm5, pd1, pd2, pd3, pd4, pd5}

Γ_h – {p, q, r, s, t, u, v, w, x, y, z} – presented in Figure 6.8.

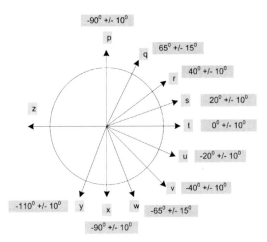

Fig. 6.8. Elements of the Γ_h set

S_h – the start symbol of the grammar,
P_h – the finite set of productions defined as follows (Figure 6.9):

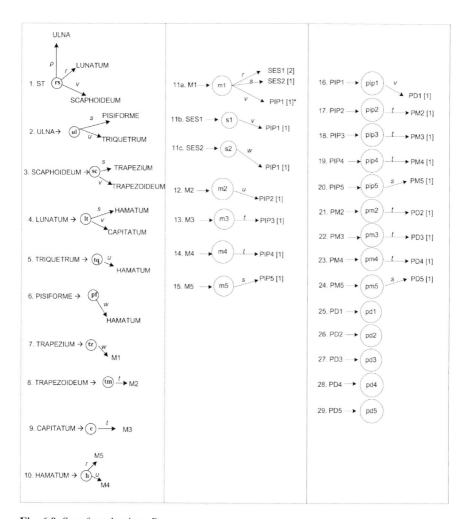

Fig. 6.9. Set of productions P_h

Figure 6.10 shows the graph of relations between individual finger, metacarpus and wrist bones.

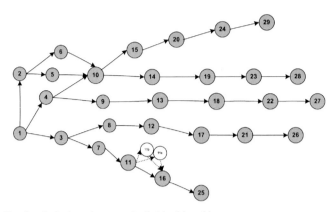

Fig. 6.10. Graph of relations between individual hand bones

Figure 6.11, in turn, shows the graph of relations between individual finger, meta-carpus and wrist bones including the angles of slopes between individual palm bones.

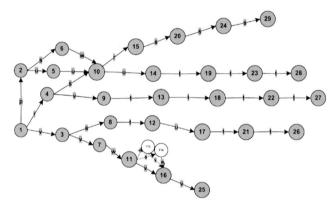

Fig. 6.11. Graph of relations between individual hand bones including the angles of slopes between them

This definition method allows the UBIAS system to start analysing image data. Selected results of its operation are illustrated in Figures 6.12-6.16, which show chosen examples of automatic image data interpretation and its semantic analysis.

The automatic analysis processes covered the following groups of pathologies observed within the hand:

- finger fractures;
- hand bone fractures;
- degenerative changes (bone fusions or bone atrophy);
- hand deformation;
- bone displacements;
- arthritis;
- tuberculosis of bones.

A cognitive data analysis system analysing lesions occurring within hand bones shown in X-ray images is an example of a cognitive system applied to the semantic analysis of image data.

Figure 6.12 shows an example of the automatic analysis of an image depicting a deformation of the radial bone.

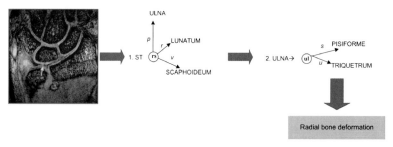

Fig. 6.12. An image data analysis by an UBIAS system to understand data showing a radial bone deformation

Figure 6.13 shows an example of the automatic analysis of an image depicting a complex of post-traumatic dislocations of wrist bones.

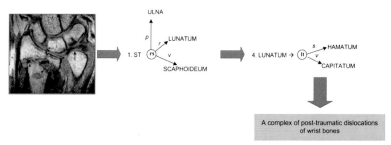

Fig. 6.13. An image data analysis by an UBIAS system to understand data showing a complex of post-traumatic dislocations of wrist bones

Figure 6.14 shows an example of the automatic analysis of an image depicting a radial bone tumour.

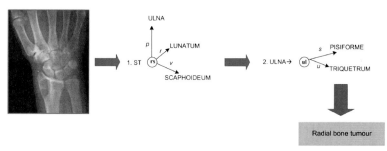

Fig. 6.14. An image data analysis by an UBIAS system to understand data showing a radial bone tumour

Figure 6.15 shows an example of the automatic analysis of an X-ray depicting rheumatoid arthritis.

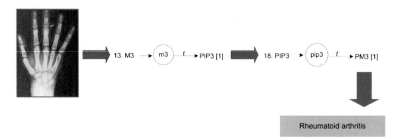

Fig. 6.15. An image data analysis by an UBIAS system to understand data showing rheumatoid arthritis

Figure 6.16 shows an example of the automatic analysis of an X-ray depicting sarcoidosis.

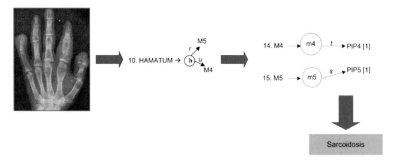

Fig. 6.16. Image data analysis by UBIAS systems to understand data showing sarcoidosis

A class of semantic analysis systems thus defined and complemented with new solutions described at the beginning of this chapter can be successfully used for identifying and verifying persons.

The biometric analysis complementing semantic analysis processes in cognitive systems proceeds as follows (Figure 6.17).

Based on a cognitive analysis hinging on the interpretation and the semantic analysis of hand bone X-rays, the information system conducts a process of understanding image data to indicate and define the lesions occurring within the hand skeleton. These lesions are stored in records linked to specific persons, and at an additional stage, the system executes a biometric analysis whose results are tied to the records of a given person. This process supports not only identifying the given individual, but in addition also:

- assigning the lesions observed in a given person to him/her;
- identifying a person based on the characteristic features of his/her hands;

- if lesions occur in a given person, he/she can be identified using the classification of hand pathologies;
- verifying a person using the biometric information and that on possible hand bone pathologies collected in the system.

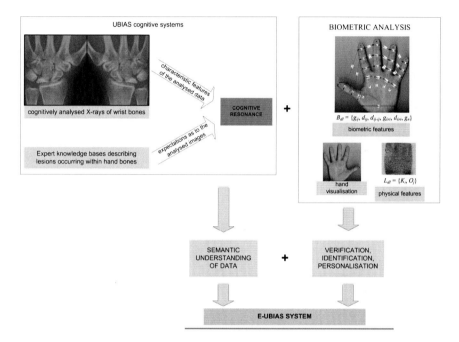

Fig. 6.17. E-UBIAS system in biometric analysis

The biometric analysis used in E-UBIAS systems offers opportunities for identifying and verifying people based on available sets of characteristic features and in addition allows the databases linked to particular individuals to be complemented with semantic characteristics. These characteristics unambiguously enable the enhanced data sets to be recognised and understood and at the same time can be used to accurately eliminate immaterial data.

Chapter 7
Cognitive Systems and Artificial Brains

Cognitive systems founded in the fields of artificial brain design and operation show the direction in which computer science is developing hand in hand with cognitive science.

Various applications of artificial and automatic solutions more and more often delight the world, but we are also increasingly wondering about the extent to which science can catch up with the reality and nature, which in an individual, unique and not fully comprehensible way creates us and the world around us. To what extent can we build an artificial brain and an artificial robot? How far can we make that robot imitate human action? And the last, possibly the most crucial question, to what extent is a robot able to imitate human skills and feelings?

If science today is capable of designing a robot which imitates human actions to some extent, will the latter also be able to smell the scents, be delighted with beauty, reflect on thoughts, or do these characteristics belong to the human kind only? Attempts are now made to apply system solutions to analyse smells and related human sensations. This work is extremely complicated, as analyses of this type are based on individual human sensations. For one person, the smells of perfume, tea, coffee, flowers, animals and the surroundings may be quite agreeable, while for another, they may be unpleasant. In addition, every group of smells, e.g. of flowers (plants), can evoke completely different feelings. Some may be delighted by the smell of evening stock, for others it is too strong and intense. So creating an ideal model which can be the foundation for designing a system analysing smell sensations may be a huge scientific challenge.

The same applies to designing a robot capable of understanding feelings and sensations. There is no ideal model for defining satisfaction, joy, anger, annoyance, friendship or even love. Every one of us defines feelings differently (sometimes even subjectively), and everyone feels such emotions differently (individually).

Building a robot that could execute tasks not as complex as analysing smells, for instance, but simple ones identical to those of humans, is now focusing around the design of cognitive machines and cognitive robots. The main idea behind such designs is the application of machine solutions, primarily cognitive ones. These solutions consist in developing cognitive systems capable of imitating the greatest possible number of human behaviours. The imitation process is based on the course of human cognition, analysis, interpretation, reasoning and thinking processes. These processes, in turn, form the foundations for designing cognitive solutions for the automatic interpretation of data and machine actions.

One program within the field of cognitive robot design was conducted in 2004-2008 by a consortium of European research institutions dealing with robotics, cognitive robots, artificial intelligence and computer visualisation. The program is known as *COGNIRON Project LAAS-CNRS*. It has led to the design of a robot that [169]:

L. Ogiela and M.R. Ogiela: Advances in Cognitive Information Systems, COSMOS 17, pp. 99–106.
springerlink.com   © Springer-Verlag Berlin Heidelberg 2012

- can conduct a multimodal conversation (a dialogue);
- can perceive orders;
- can recognise orders;
- can behave like a human;
- can impersonate human interactions (reactions);
- can learn;
- is aware of the space, analyses the spatial situation;
- recognises multimodal situations and is aware of their occurrence;
- can execute its (simple) plans.

The Cogniron project was one of many implemented in the field of cognitive robot building. Other work of this type was also undertaken by:

- The United States Naval Research Laboratory and NCARAI Intelligent Systems to develop a robot under the charming name of Octavia [189].

Octavia is a cognitive robot that recognises a given situation, analyses it as well as its surroundings, can make human gestures, is agile, mobile and humanoid. The robot's face shows expressions, which have never before been associated with robots. In addition, Octavia can learn new solutions based on situations that take place (and are new for it). The movements of the mouth and eyes are very endearing, because although made by a robot, they seem to reflect expressions on a human face in a very human way. Figure 7.1 shows the cognitive robot named Octavia.

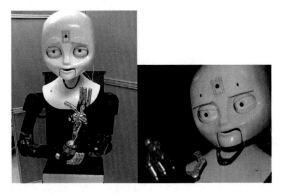

Fig. 7.1. The Octavia Robot. Source: [189]

- The Vanderbilt University School of Engineering, Cognitive Robotics Laboratory, when creating the ISAC humanoid robot design [161].

ISAC is a robot that recognises the voice of a person giving it orders and constitutes an interactive solution maintaining the human-robot relation, understands the voice and can answer. It remembers earlier situations, both those from a moment and a longer time ago, so it has layers of short-term memory and some elements of long-term one from which it can recall memorized situations. ISAC operates using various memory structures, learning algorithms, observation mechanisms and methods of decision-making adapted to its structure. Figure 7.2 shows ISAC, the humanoid robot.

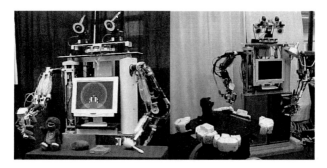

Fig. 7.2. The ISAC humanoid robot. Source: [161]

- DARPA/NASA-JSC in their work on Robonaut [193].

The robot called Robonaut is an extremely agile humanoid robot built by the NASA L.B. Johnson Space Center in Houston, Texas. Robonaut was built for a specific purpose – to help people in their work – to perform human activities associated with research conducted in outer space. Due to its jobs, Robonaut is (from the waist up) a copy of an American astronaut, and below is a vehicle allowing the whole to quickly move in difficult conditions over various surfaces.

The newer version of the robot is Robonaut 2, faster and more agile than its predecessor. It is characterised by greater perception (as it has deeper and wider sensory perception). Figure 7.3 shows both versions of the humanoid robot.

Fig. 7.3. The Robonaut and Robonaut 2 humanoid robots. Source: [193]

- The robot named Simon constructed at the Georgia Institute of Technology [177].

Simon is a solution for cognitively analysing its surroundings, situations and for interpreting data. It has been presented at demonstrations and on the Georgia Tech website to show how it analyses, for instance, the colours of various objects, which, after their correct recognition, it places in containers of the same colour. Such solutions illustrate how a robot can assign an interval of the colour spectrum to a given model, as every container is of a distinctive colour, while the objects recognised, although they are of the given colour, have various colour intensities. In addition, during the recognition analysis conducted by Simon, the robot's ears shine the recognised colour, which additionally allows an observer of the robot to assess that the colour recognition it has conducted is right (Figure 7.4).

Fig. 7.4. The Simon robot. Source: [177]

- The Toyota robot playing a musical instrument presented in 2005 by the Toyota Motor Corporation (TMC) in walking and driving versions [195].

The first version of the robot moves in the traditional way, by walking, and can play a trumpet of beautifully tone. The robot's arm and leg movements are precise and efficient. The second version of the Toyota robot moves using a wheeled mechanism and just as the walking Toyota plays beautiful notes on its trumpet. They make gestures (e.g. raising the hand, waving) and movements (e.g. of the trumpet, movements of the robot's 'body') while recognising the action they are to perform (e.g. playing the work of music). Because the bodies of these robots have been fully clad, they look very nice, unlike the majority of robots of this type whose 'bodies' are incomplete (Figure 7.5.)

Fig. 7.5. The Toyota robot. Source: [195]

Another solution from this group of musical robots plays the violin. It can perform the same functions as the trumpet-playing Toyota (Figure 7.6).

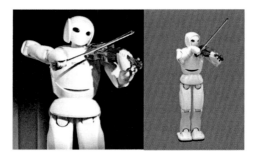

Fig. 7.6. The Toyota robot. Source: [195]

The Japanese really love to develop automatic robots performing varied tasks. It is enough to mention even just the solutions aimed at raising the quality of life by doing various household chores. An example of such a robot is shown in Figure 7.7 – a robot cleaning the office, and now widely known iRobots that clean flat surfaces (sweeping, polishing).

Fig. 7.7. An office/household robot, an iRobot. Source: [195]

Figure 7.8 shows a Japanese proposal of a folk robot which can act as a museum guide wearing a beautiful national costume.

Fig. 7.8. A museum robot. Source: [195]

Regardless of the type of the few solutions presented here from among so many, cognitive robots must be capable of performing activities which frequently require a lot of precision. If we are to develop a robot used in production processes, in medicine during complex operations, in the military for locating a target to be analysed (not excluding tactical operations), in outer space work or in many other applications, there is no doubt of the precision and accuracy of action indispensable in these cases. For a robot to have such characteristics, it must analyse the situation correctly, or more precisely, it has to correctly identify the purpose of its action (of its analysis). These activities are characteristic for cognitive robots which not only execute simple actions, but also, or primarily, can understand an instruction, the given situation or their surroundings. Activities taken by a cognitive robot are shown in Figure 7.9.

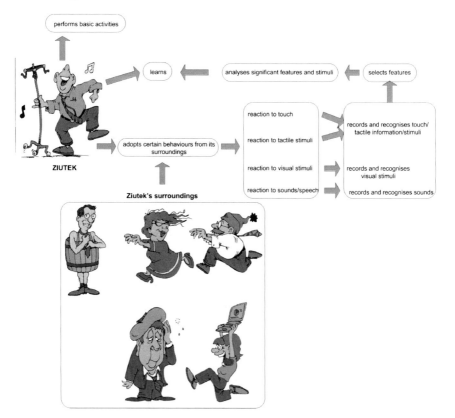

Fig. 7.9. A cognitive robot architecture

Figure 7.9 shows the architecture of a cognitive robot named Ziutek, the flows of information reaching Ziutek as well as the method of its perception and analysis. The robot executes basic actions including:

- moving (can walk, can ride);
- head and hand movements;
- facial expressions (narrowing eyelids, movements of the mouth);
- gestures (satisfaction, dissatisfaction, tiredness);

- grasping tools, objects;
- answering questions asked;
- executing activities like handing a book, pouring water into a mug, lifting an object from a given place etc.

Apart from the basic actions which can be freely defined for a cognitive robot depending on its intended function , the robot must also be able to perform other tasks. The most important one is learning. This process is based on the robot's relations with its environment which sends certain signals to it. For the robot, these signals are a certain type of behaviours which it perceives as new or as known. Behaviours consisting in waving a hand in greeting or farewell to the robot acts as visual stimuli for it. These stimuli are recorded in the robot's artificial brain and then processed and analysed. As a result of this process, Ziutek the robot may wave a hand to us, smile or nod its head. The robot also records other types of stimuli, which include touch stimuli (understood as touching the robot directly), tactile stimuli (e.g. wind, a draught in a room), aural stimuli including sounds (e.g. of music from a player) and speech (distinguished as a separate aural stimulus). The robots registers all these stimuli and cognitively analyses them. If the recognition process runs correctly, the robot takes 'responsive' action showing that the situation has been analysed appropriately. If not, the robot may have problems with understanding the analysed situation. It must then try to learn the new situation, new solutions, and these activities constitute a type of adaptive action for the robot. The learning process consists in selecting significant characteristics of the analysis process executed, which enrich the knowledge base held by the robot.

The actions taken by the robot are controlled. Control is founded on analysing cognitive processes taking place in the robot's 'brain'. Every stimulus sent/ perceived/received initiates sensors which cause certain reactions. These reactions are the resultant of much varied information and many actions, including experience associated with similar situations, information stored in memory layers, information of similar meaning, information worth trying to adapt to the analysed situation and defined (programmed) executive functions. All this information, functions and experience are interconnected and their consistency, i.e. the lack of information about interference occurring, represents a control stage. It is at this stage that if any inconsistencies are detected between the elements of the reaction caused, the robot's action is interrupted or it is irrational (contrary to our expectations). If all the above elements are consistent, the stimulating reaction is caused and ends in taking a specific action or operation (Figure 7.10).

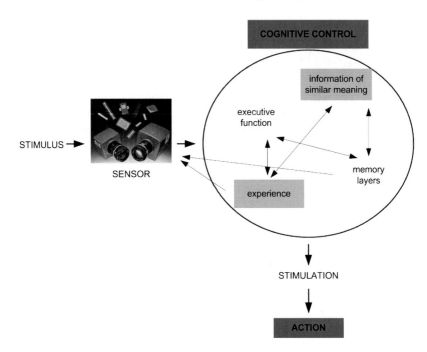

Fig. 7.10. The cognitive control process

Cognitive robots are increasingly frequently and readily designed for the purposes of semantic analyses, defence, improving production processes, logistics, transport, medicine and many other fields of science and everyday life in which mobile operation can no longer be imagined without this type of solutions.

Yet cognitive robots, seen to develop very intensely and dynamically towards avatars – machines imitating people, their behaviour and action – still seem to have a long way to go. This is because an artificial copy of a human cannot be built, but it is possible to build a robot equipped with a set of features characteristic for human behaviour. Consequently, the notion of an avatar seems to significantly exceed the cognitive, design and technical capabilities, whereas the concept of a cognitive robot fits the work on developing an artificial brain quite accurately.

Chapter 8
Summary

Cognitive informatics developed in the direction of broader and broader alliances between technical computer science and cognitive science represents a very interesting and promising field of science. Combining the whole range of solutions and mechanisms that can be developed by cognitive science with computer science opens wider and greater application opportunities. The solutions shown for problems presented herein also demonstrate further development opportunities for the proposed solutions due to the constant development of cognitive informatics.

Cognitive data analysis and interpretation models make it possible to extract new, extremely significant information – semantic information – from the analysed data sets. It is this type of information which makes its deeper meaning-based interpretation possible. Semantic interpretation uncovers deep layers of information for the person conducting the analysis process. They thus represent an extension of traditional data interpretation attempts whose end result was recognising the analysed object, feature, information or data. Cognitive informatics based on semantic data interpretation processes allows meaning to be extracted from sets of this data, which meaning guides the automatic understanding process. This process is executed by imitating human processes of perception, cognition, analysis and consequently, understanding, which can form the foundation for reasoning and projecting changes that may occur in the future. Changes of the analysed phenomenon are an inseparable element of cognitive data analysis processes, as the projection processes enable the identification of possible improvements, or worse, deterioration. In cognitive analysis of medical images of lesions, this type of projection is of great importance for the patient's life and health. Similarly, conditions describing the economic standing of an enterprise or the financial premises for making a given investment are also of great importance. They may not be about life and death, but for investors, financial advisors, economists and finally the economy they may bring a time of financial downturn or prosperity. Cognitive solutions show how significant the stages of understanding the analysed data are. It is during this process that cognitive systems take decisions to correctly classify the analysed situation and the consequences of its occurrence, or if the analysed data is not understood, they attempt to include the new solutions in the knowledge bases used for the cognitive data understanding process. This stage, as an element of the system learning process, allows the system to constantly improve and thus enables it to understand newer and newer situations.

Cognitive informatics is open to the semantic analysis of more than one type of data. It has also offered opportunities for identification analyses, called the personalisation. Analysis methods based on cognitive and biometric mechanisms of personal identification and verification made it possible to combine semantic interpretation techniques with personal authentication and identification technologies. This combination gave birth to E-UBIAS systems which play a major role in

L. Ogiela and M.R. Ogiela: Advances in Cognitive Information Systems, COSMOS 17, pp. 107–108.
springerlink.com

correctly identifying people. This role consists not just in correctly recognising the individual, but primarily makes it possible to build bases using individual features of the individual, enhanced with their semantic aspects. It is those semantic parameters that give the certainty that the given person (if they are characterised by any lesion, e.g. in their hand skeleton) will be correctly identified using their individual biometric features and verified using lesions. Thus the wrong identification of a given individual becomes highly improbable. If the system were to incorrectly assign biometric features to a person who, during the personal verification process, has to undergo a two-stage identification analysis, this result will still be subject to the second stage of the semantic analysis. Thus if the given person has some lesions, they allow this person to be correctly identified. If he/she has no lesions, the analysis process obviously ends at the first of the above stages. This is why, in their operation, E-UBIAS systems combine the analysis of biometric features of a given person and the cognitive analysis of possible lesions present.

Contemporary systems of cognitive data analysis are also used to build an artificial brain and automatic 'imitations' of Man, called cognitive robots. Their variety, numerous solutions and improvement attempts show how many of us are fascinated by this type of work. This is probably due not just to the splendid opportunities for using cognitive robots, but primarily to the great satisfaction with being able to build a real (though machine and automatic) solution. If we just look at the examples of cognitive robots cited here (selected from a much more numerous group), we get the inescapable feeling that the 'faces' of these robots show the great satisfaction they gave to their authors. Cognitive robots are solutions with bright future before them, because their wide range of application can still be stretched in many different directions. They also represent the 'hope' of the world of science that we will at least partly be able to build an artificial brain, not as a replica of the human one, as this is impossible, but as its most complete version reflecting its ideal to the greatest possible degree. This is a very difficult job and none of us can answer the question: *How well can an artificial brain imitate the human one and to what extent?*

References

1. Albus, J.S., Meystel, A.M.: Engineering of Mind – An Introduction to the Science of Intelligent Systems. A Wiley-Interscience Publication John Wiley & Sons Inc. (2001)
2. Anderson, J.R.: The Architecture of Cognition. Harvard Univ. Press, Cambridge (1983)
3. Anderson, J.R.: ICCI 2005, Proc. 7th International Conference on Cognitive Informatics, ICCI 2008. IEEE CS 17 Press, Stanford University, CA (2005)
4. Aristotle: The Basic Works of Aristotle, by Richard McKeon. Random House Inc. (1941)
5. Atkinson, R.C., Shiffrin, R.M.: Human memory, A proposed system and its control processes. In: Spence, K.W., Spence, J.T. (eds.) The Psychology of Learning and Motivation: Advances in Research and Theory, pp. 89–195. Academic Press, New York (1968)
6. Bechtel, W., Abrahamsen, A., Graham, G.: The live of cognitive science. In: Bechtel, W., Graham, G. (eds.) A Companion of Cognitive Science, pp. 1–104. Blackwell Publishers, UK (1998)
7. Bell, D.A.: Information Theory. Pitman, London (1953)
8. Bender, E.A.: Mathematical Methods in Artificial Intelligence. IEEE CS Press, Los Alamitos (1996)
9. Bernstein, L., Wild, J.: Analysis of Financial Statements, 5th edn., Amazon (1999)
10. Bocheński, I.M.: Ancient Formal Logic. North-Holland Publishing Company, Amsterdam (1951)
11. Borod, J.C.: Interhemispheric and Intrahemispheric Control of Emotion: A Focus on Unilateral Brain Damage. Journal of Consulting and Clinical Psychology 60, 339–348 (1992)
12. Branquinho, J. (ed.): The Foundations of Cognitive Science. Clarendon Press, Oxford (2001)
13. Brejl, M., Sonka, M.: Medical image segmentation: Automated design of border detection criteria from examples. Journal of Electronic Imaging 8(1), 54–64 (1999)
14. Brooks, R.A.: A robust layered control system for a mobile robot. IEEE Journal of Robotics and Automation RA-2, 14–23 (1986)
15. Burgener, F.A., Kormano, M.: Bone and Joint Disorders, Conventional Radiologic Differential Diagnosis. Thieme, Stuttgart-New York (1997)
16. Burgener, F.A., Meyers, S.P., Tan, R.K., Zaunbauer: Differential Diagnosis in Magnetic Resonance Imaging. Georg Thieme Verlag (2002)
17. Cellary, W.: Multi-version Serializability and Concurrency Control. In: Tamer, M., Liu, L. (eds.) Encyclopedia of Database Systems (in 5 volumes), pp. 208–211. Springer, Heidelberg (2009)
18. Chan, C., Kinsner, W., Wang, Y., Miller, D.M. (eds.): Proc. 3rd IEEE International Conference on Cognitive Informatics (ICCI 2004). IEEE CS Press, Victoria (2004)
19. Chomsky, N.: Language and Mind. Harcourt Brace Javanovich, New York (1972)

20. Chomsky, N.: Language and Problems of Knowledge. The Managua Lectures. MIT Press, Cambridge (1988)
21. Cios, K.J., Kurgan, L.A.: Trends in Data Mining and Knowledge Discovery. In: Pal, N.R., Jain, L.C. (eds.) Advanced Techniques in Knowledge Discovery and Data Mining, pp. 1–26. Springer, Heidelberg (2005)
22. Cios, K., Pedrycz, W., Swiniarski, R., Kurgan, L.A.: Data Mining: A Knowledge Discovery Approach. Springer, Heidelberg (2007)
23. Cleeremans, A.: Principles for implicit learning. In: Berry, D.C. (ed.) How Implicit is Implicit Learning?, pp. 195–234. Oxford University Press, London UK (1997)
24. Cohen, H., Lefebvre, C. (eds.): Handbook of Categorization in Cognitive Science. Elsevier, The Netherlands (2005)
25. Davidson, R.: Prolegomenon to Emotion: Gleanings from Neuropsychology. Cognition and Emotion 6, 245–268 (1992)
26. Davis, L.S. (ed.): Foundations of Image Understanding. Kluwer Academic Publishers (2001)
27. Dennett, D.C.: Consciousness Explained. Little Brown and Co., Boston (1991)
28. Dreyfus, H.: What Computers Still Can't Do. MIT Press, Cambridge (1992)
29. Driscoll, M.: Psychology of Learning for Instruction. Allyn and Bacon (1991)
30. Duda, R.O., Hart, P.E., Stork, D.G.: Pattern Classification, 2nd edn. A Wiley-Interscience Publication, John Wiley & Sons, Inc. (2001)
31. Edelman, S.: Representation and Recognition in Vision. MIT Press, Cambridge (1999)
32. Edwards, W.: The theory of decision making. Psychological Bulletin 51, 380–417 (1953)
33. Ekman, P.: Facial Expressions and Emotion. American Psychologist 48, 348–392 (1993)
34. Evans, T.G.: A heuristic program to solve geometric-analogy problems. In: Minsky, M. (ed.) Semantic Information Processing, pp. 271–353. MIT Press, Cambridge (1968)
35. Falkenhainer, B., Forbus, K.D., Gentner, D.: The structure mapping engine: Algorithm and examples. Artificial Intelligence 41, 1–63 (1989)
36. Flasiński, M.: Inference of Parsable Graph Grammars for Syntactic Pattern Recognition. In: Fundamenta Informaticae, vol. 80, pp. 379–413. IOS Press, Amsterdam (2007)
37. Fodor, J.A.: The modularity of mind. MIT Press, Cambridge (1983)
38. Gabrieli, J.D.E.: Cognitive Neuroscience of Human Memory. Annual Review of Psychology 49, 87–115 (1998)
39. Gigerenzer, G., Todd, P.M. (eds.): The ABC Research Group: Simple heuristics that make us smart. Oxford University Press, Oxford (1999)
40. Good, I.J.: Speculations concerning the first ultra intelligent machine. In: Alt, F.L., Rubinoff, M. (eds.) Advences in Computers 6, pp. 31–88. Academic Press, New York
41. Gray, P.: Psychology, 2nd edn. Worth Publishers, Inc., New York (1994)
42. Hebb, D.O.: The Organization of Behavior. Wiley, New York (1949)
43. Hebb, D.O.: A Textbook of Psychology. W. B. Sanders, Philadelphia (1958)
44. Helmoholtz, H. (ed.): Epistemological writings. Boston Studies in the Philosophy of Science, vol. 79. Reidel, Dordrecht (1977)
45. Huang, H.K., Kangarloo, H., Cho, P.S., Taira, R.K., Ho, B.K.T., Chan, K.K.: Planning a totally digital radiology department. AJR 154, 635–639 (1990)

46. Huang, H.K.: Picture Archiving and Communication Systems in Biomedical Imaging. VCH Publishers, Inc. (1996)
47. Hubel, D.: Eye, Brain and Vision. W. H. Freeman, New York (1988)
48. Hubel, D., Wiesel, T.: Receptive Fields, Binocular Interaction, and Functional Architecture in the Cat's Visual Cortex. Journal of Physiology 160, 106–154 (1962)
49. IBM: Autonomic Computing White Paper: An Architectural Blueprint for Autonomic Computing, 4th edn., June 2006, pp. 1-37 (2006)
50. Jurek, J.: Recent developments of the syntactic pattern recognition model based on quasi-context sensitive languages. Pattern Recognition Letters 2(26), 1011–1018 (2005)
51. Kagan, J.: Unstable Ideas: Temperament, Cognition and Self. Harvard University Press, Cambridge, Mass. (1989)
52. Kahneman, D., Tversky, A.: Prospect theory: An analysis of decision under risk. Econometrica 47, 263–293 (1979)
53. Kamp, H., Reyle, U.: From Discourse to Logic. Kluwer, Dordrecht (1993)
54. Kephart, J., Chess, D.: The Vision of Autonomic Computing. IEEE Computer 26(1), 41–50 (2003)
55. Kickhard, M., Terveen, L.: Foundational Issues in Artificial Intelligence and Cognitive Science. Elsevier, Amsterdam (1996)
56. Kihlstrom, J.F.: The Cognitive Unconscious. Science 237, 1445–1452 (1987)
57. Kinsner, W., Zhang, D., Wang, Y., Tsai, J. (eds.): Proc. 4th IEEE International Conference on Cognitive Informatics (ICCI 2005). IEEE CS Press, Irvine (2005)
58. Kinsner, W.: Towards Cognitive Machines: Multiscale Measures and Analysis, Keynote. In: Proc. 5th International Conference on Cognitive Informatics (ICCI 2006), pp. 8–14. IEEE CS Press, Beijing (2006)
59. Latombe, J.C.: Probabilistic Roadmaps: A Motion Planning Approach Based on Active Learning, Keynote. In: Proc. 5th International Conference on Cognitive Informatics (ICCI 2006), p. 1. IEEE CS Press, Beijing (2006)
60. Laudon, K.C., Laudon, J.P.: Management Information Systems – Managing the Digital Firm, 7th edn. Prentice-Hall International, Inc. (2002)
61. Lazarus, R.S.: Emotion and Adaptation. Oxford University Press, New York (1991)
62. Leahey, T.H.: A History of Psychology: Main Currents in Psychological Thoughts, 4th edn. Prentice Hall, Upper Saddle River (1997)
63. Lehrl, S., Fischer, B.: The basic parameters of human information processing: their role in determination of intelligence. Personality and Individual Differences 9, 883–896 (1988)
64. Leibniz, G.W.: Dissertatio de arte combinatoria. Leibnizens Mathematische Schriften 5. Georg Olms, Hildesheim (1666)
65. Lenat, D.B.: Cyc: A large-scale investment in knowledge infrastructure. Communications of the ACM 38, 33–38 (1995)
66. Llull, R.: Logica Nova. In: Bonner, A. (ed.) Nova Edicio de Les Obres de Ramon Llull 1303, vol. 4, Palma de Mallorca (1998)
67. Mandler, G.: Mind and Body: Psychology of Emotion and Stress. Norton, New York (1984)
68. Marieb, E.N.: Human Anatomy and Physiology, 2nd edn. The Benjamin/Cummings Publishing Co., Inc., Redwood City (1992)
69. Marr, D.: Vision. Freeman, San Francisco (1982)
70. Masterman, M.: Translation. Proceedings of the Aristotelian Society, pp. 169–216 (1961)

71. McCarthy, J.: Recursive Functions of Symbolic Expressions. Communications of the Association for Computing Machinery 3 (1960)

72. McClosey, M., Glucksberg, S.: Natural categories: Well-defined or fuzzy sets? Memory and Cognition 6, 462–472 (1978)

73. Meystel, A.M., Albus, J.S.: Intelligent Systems – Architecture, Design, and Control. A Wiley-Interscience Publication, John Wiley & Sons, Inc., Canada (2002)

74. Mill, J.S.: A System of Logic. Longmans, London (1865)

75. Miller, E.K.: Cognitive Control. Fundamental of Brain and Mind Lecture Series, Mass. Institute of Technology, June 11-13 (2003)

76. Minsky, M.: Semantic Information Processing. MIT Press, Cambridge (1968)

77. Minsky, M.: A framework for representing knowledge. In: Winston, P.H. (ed.) The Psychology of Computer Vision. McGraw-Hill, New York (1975)

78. Minsky, M.: The Society of Mind. Simon & Schuster, New York (1987)

79. Murphy, G.L., Medin, D.L.: The Role of Theories in Conceptual Coherence. Psychological Review 92, 289–315 (1985)

80. Narayanan, S.: Reasoning about actions in narrative understanding. In: Proceedings of the International Joint Conference on Artificial Intelligence, pp. 350–358 (1999)

81. Newell, A.: Unified Theories of Cognition. Harvard University Press, Cambridge (1990)

82. Ogiela, L.: Cognitive Systems for Medical Pattern Understanding and Diagnosis. In: Lovrek, I., Howlett, R.J., Jain, L.C. (eds.) KES 2008, Part I. LNCS (LNAI), vol. 5177, pp. 394–400. Springer, Heidelberg (2008)

83. Ogiela, L.: Modelling of Cognitive Processes for Computer Image Interpretation. In: Al-Dabass, D., Nagar, A., Tawfik, H., Abraham, A., Zobel, R. (eds.) EMS 2008 European Modelling Symposium, Second UKSIM European Symposium on Computer Modeling and Simulation, Liverpool, United Kingdom, September, 8-10, pp. 209–213 (2008)

84. Ogiela, L.: Syntactic Approach to Cognitive Interpretation of Medical Patterns. In: Xiong, C.-H., Liu, H., Huang, Y., Xiong, Y.L. (eds.) ICIRA 2008. LNCS (LNAI), vol. 5314, pp. 456–462. Springer, Heidelberg (2008)

85. Ogiela, L.: Cognitive Computational Intelligence in Medical Pattern Semantic Understanding. In: 4th International Conference on Natural Computation ICNC 2008, Jinan, China, October 18-20, vol. 6, pp. 245–247 (2008)

86. Ogiela, L.: Innovation Approach to Cognitive Medical Image Interpretation. In: Innovation 2008, 5th International Conference on Innovations in Information Technology, Al Ain, United Arab Emirates, December 16-18, pp. 722–726 (2008)

87. Ogiela, L.: UBIAS Systems for the Cognitive Interpretation and Analysis of Medical Images. Opto-Electronics Review 17(2), 166–179 (2009)

88. Ogiela, L.: Computational Intelligence in Cognitive Healthcare Information Systems. In: Bichindaritz, I., Vaidya, S., Jain, A., Jain, L.C. (eds.) Computational Intelligence in Healthcare 4. SCI, vol. 309, pp. 347–369. Springer, Heidelberg (2010)

89. Ogiela, L.: Cognitive Informatics in Automatic Pattern Understanding and Cognitive Information Systems. In: Wang, Y., Zhang, D., Kinsner, W. (eds.) Advances in Cognitive Informatics and Cognitive Computing. SCI, vol. 323, pp. 209–226. Springer, Heidelberg (2010)

90. Ogiela, L.: Semantic Analysis Processes in UBIAS Systems for Cognitive Data Analysis. Special Issue on Advances in Context, Cognitive, and Secure Computing in CAMWA Journal (in press 2011)

91. Ogiela, L.: Pattern Classifications in Cognitive Informatics. In: Ogiela, M., Jain, L. (eds.) Computational Intelligence Paradigms in Advanced Pattern Classification. SCI. Springer, Heidelberg (in Press, 2011)

92. Ogiela, L., Ogiela, M.R.: Cognitive Techniques in Visual Data Interpretation. SCI, vol. 228. Springer, Heidelberg (2009)

93. Ogiela, L., Ogiela, M.R., Tadeusiewicz, R.: Mathematical Linguistic in Cognitive Medical Image Interpretation Systems. Journal of Mathematical Imaging and Vision 34, 328–340 (2009)

94. Ogiela, L., Tadeusiewicz, R., Ogiela, M.R.: Cognitive Analysis in Diagnostic DSS-Type IT Systems. In: Rutkowski, L., Tadeusiewicz, R., Zadeh, L.A., Żurada, J.M. (eds.) ICAISC 2006. LNCS (LNAI), vol. 4029, pp. 962–971. Springer, Heidelberg (2006)

95. Ogiela, L., Tadeusiewicz, R., Ogiela, M.R.: Cognitive Approach to Visual Data Interpretation in Medical Information and Recognition Systems. In: Zheng, N., Jiang, X., Lan, X. (eds.) IWICPAS 2006. LNCS, vol. 4153, pp. 244–250. Springer, Heidelberg (2006)

96. Ogiela, L., Tadeusiewicz, R., Ogiela, M.R.: Graph-Based Structural Data Mining In Cognitive Pattern Interpretation. In: Sakakibara, Y., Kobayashi, S., Sato, K., Nishino, T., Tomita, E. (eds.) ICGI 2006. LNCS (LNAI), vol. 4201, pp. 349–350. Springer, Heidelberg (2006)

97. Ogiela, L., Tadeusiewicz, R., Ogiela, M.R.: Cognitive Computing in Intelligent Medical Pattern Recognition Systems. In: De-Shuang, H., Kang, L., Irwin, G.W. (eds.) Intelligent Control and Automation – International Conference on Intelligent Computing, ICIC 2006, Kunming, China, August 16-19. LNCIS, vol. 344, pp. 851–856. Springer, Heidelberg (2006)

98. Ogiela, L., Tadeusiewicz, R., Ogiela, M.R.: Cognitive Categorization in Modeling Decision and Pattern Understanding. In: Torra, V., Narukawa, Y., Yoshida, Y. (eds.) MDAI 2007. LNCS (LNAI), vol. 4617, pp. 69–75 (2007)

99. Ogiela, L., Tadeusiewicz, R., Ogiela, M.R.: Cognitive Informatics In Automatic Pattern Understanding. In: Hang, D., Wang, Y., Kinsner, W. (eds.) Proceedings of the Sixth IEEE International Conference on Cognitive Informatics, ICCI 2007, Lake Tahoe, CA, USA, August 6-8, pp. 79–84 (2007)

100. Ogiela, L., Tadeusiewicz, R., Ogiela, M.R.: Cognitive Computing In Analysis of 2D/3D Medical Images. In: The 2007 International Conference on Intelligent Pervasive Computing – IPC 2007, October 11th-13th, pp. 15–18. IEEE Computer Society, Jeju Island (2007)

101. Ogiela, L., Tadeusiewicz, R., Ogiela, M.R.: Cognitive Linguistic Categorization for Medical Multi-dimensional Pattern Understanding. In: ACCV 2007 Workshop on Multi-Dimensional and Multi-View Image Processing, Tokyo, Japan, November 18-22, pp. 150–156 (2007)

102. Ogiela, L., Tadeusiewicz, R., Ogiela, M.R.: Cognitive techniques in medical information systems. Computers In Biology and Medicine 38, 502–507 (2008)

103. Ogiela, L., Tadeusiewicz, R., Ogiela, M.R.: Cognitive Approach to Medical Image Semantics Description and Interpretation. In: INFOS 2008, The 6th International Conference on Informatics and Systems, March 27-29, pp. HBI-1–HBI-5 (2008)

104. Ogiela, L., Tadeusiewicz, R., Ogiela, M.R.: Cognitive Modeling in Medical Pattern Semantic Understanding. In: The 2nd International Conference on Multimedia and Ubiquitous Engineering MUE 2008, Busan, Korea, April 24-26, pp. 15–18 (2008)

105. Ogiela, L., Tadeusiewicz, R., Ogiela, M.R.: Al-Cognitive Description in Visual Pattern Mining and Retrieval. In: Second Asia Modeling & Simulation AMS, Kuala Lumpur, Malaysia, May 13-15, pp. 885–889 (2008)

106. Ogiela, L., Tadeusiewicz, R., Ogiela, M.R.: Cognitive Modeling in Computational Intelligence Methods for Medical Pattern Semantic Categorization and Understanding. In: Proceedings of the Fourth IASTED International Conference Advances in Computer Science and Technology (ACST 2008), Malaysia, Langkawi, April 2-4, pp. 368–371 (2008)

107. Ogiela, L., Tadeusiewicz, R., Ogiela, M.R.: Cognitive Categorization in Medical Structures Modeling and Image Understanding. In: Li, D., Deng, G. (eds.) International Congress on Image and Signal Processing, CISP 2008, Sanya, Hainan, China, May 27-30, pp. 560–564 (2008)

108. Ogiela, L., Tadeusiewicz, R., Ogiela, M.R.: Cognitive Approach to Medical Pattern Recognition, Structure Modeling and Image Understanding. In: Peng, Y., Zhang, Y. (eds.) First International Conference on BioMedical Engineering and Informatics, BMEI 2008, Sanya, Hainan, China, May 27-30, pp. 33–37 (2008)

109. Ogiela L., Tadeusiewicz R., Ogiela M.R.: Cognitive Methods in Medical Image Analysis and Interpretation. In: The 4th International Workshop on Medical Image and Augmented Reality, MIAR 2008, The University of Tokyo, Tokyo, Japan, August 1-2 (2008)

110. Ogiela, L., Tadeusiewicz, R., Ogiela, M.R.: Cognitive Categorizing in UBIAS Intelligent Medical Information Systems. In: Sordo, M., Vaidya, S., Jain, L.C. (eds.) Studies in Computational Intelligence (SCI) 107, Advanced Computational Intelligence Paradigms in Healthcare 3, pp. 75–94. Springer, Heidelberg (2008)

111. Ogiela, L., Tadeusiewicz, R., Ogiela, M.R.: Understanding Based Managing Support Systems: The Future of Information Systems. In: Parthasarathy, S. (ed.) Enterprise Information Systems and Implementing IT Infrastructures – Challenges and Issues, Business Science Reference, pp. 91–102. IGI Global, Hershey, New York (2010)

112. Ogiela, M.R., Tadeusiewicz, R.: Modern Computational Intelligence Methods for the Interpretation of Medical Images. Springer, Heidelberg (2008)

113. Ogiela, M.R., Tadeusiewicz, R., Ogiela, L.: Syntactic Pattern Analysis in Visual Signal Processing and Image Understanding. In: The International Conference on Fundamentals of Electronic, Communications and Computer Science – ICFS 2002, Tokyo, Japan, March 27-28, pp. 13-10 – 13-14 (2002)

114. Ogiela, M.R., Tadeusiewicz, R., Ogiela, L.: Intelligent Semantic Information Retrieval In Medical Pattern Cognitive Analysis. In: Gervasi, O., Gavrilova, M.L., Kumar, V., Laganá, A., Lee, H.P., Mun, Y., Taniar, D., Tan, C.J.K. (eds.) ICCSA 2005. LNCS, vol. 3483, pp. 852–857. Springer, Heidelberg (2005)

115. Ogiela, M.R., Tadeusiewicz, R., Ogiela, L.: Cognitive Visio Techniques in Medical Image Processing and Analysis. In: Healthcom 2005, 7th International Workshop on Enterprise Networking and Computing in Healthcare Industry, Busan, Korea, June 23-25, pp. 120–123 (2005)

116. Ogiela, M.R., Tadeusiewicz, R., Ogiela, L.: Graph image language techniques supporting radiological, hand image interpretations. Computer Vision and Image Understanding 103, 112–120 (2006)

117. Ogiela, M.R., Tadeusiewicz, R., Ogiela, L.: Image languages in intelligent radiological palm diagnostics. Pattern Recognition 39, 2157–2165 (2006)

118. Patel, D., Patel, S., Wang, Y. (eds.): Cognitive Informatics: Proc. 2nd IEEE International Conference (ICCI 2003). IEEE CS Press, London (2003)

119. Pedrycz, W. (ed.): Granular Computing: An Emerging Paradigm. Physica-Verlag, Heidelberg (2001)
120. Pinel, J.P.J.: Biopsychology, 3rd edn. Allyn and Bacon, Needham Heights (1997)
121. Rapcsak, S.Z., Comer, J.F., Rubens, A.E.: Anomia for Facial Expressions: Neuropsychological Mechanisms and Anatomical Correlates. Brain and Language 45, 233–252 (1993)
122. Reisberg, D.: Cognition, 2nd edn. Exploring the science of the mind. W.W. Norton & Company, Inc. (2001)
123. Rutkowski, L.: Computational Intelligence. Methods and Techniques. Springer, Heidelberg (2008)
124. Schachter, S., Singer, J.: Cognitive, Social and Physiological Determinants of Emotional State. Psychological Review 63, 379–399 (1962)
125. Schank, R.C.: Conceptual Information Processing. North-Holland Publishing Company, Amsterdam (1975)
126. Searle, J.R.: Umysł, mózg i nauka, Wyd. PWN, Warszawa (1995) (in polish)
127. Shannon, C.E.: A Mathematical Theory of Communication. Bell System Technical Journal 27, 379–423 and 623-656 (1948)
128. Simon, H.A.: Administrative behavior. Macmillan, New York (1947)
129. Simon, H.A.: Information-processing theory of human problem solving. In: Estes, W.K. (ed.) Handbook of Learning and Cognitive Processes: Human Information. Lawrence Erlbaum, Oxford (1978)
130. Sloane, P., MacHale, D.: Perplexing Lateral Thinking Puzzles, Sterling, New York (1997)
131. Smith, R.E.: Psychology. West Publishing Co., St. Paul, MN (1993)
132. Solso, R.L. (ed.): Mind and Brain Science in the 21st Century. The MIT Press, Cambridge (1999)
133. Sternberg, R.J.: Search of the Human Mind, 2nd edn. Harcourt Brace & Co., Orlando (1998)
134. Tadeusiewicz, R., Ogiela, L.: Selected Cognitive Categorization Systems. In: Rutkowski, L., Tadeusiewicz, R., Zadeh, L.A., Zurada, J.M. (eds.) ICAISC 2008. LNCS (LNAI), vol. 5097, pp. 1127–1136. Springer, Heidelberg (2008)
135. Tadeusiewicz, R., Ogiela, L., Ogiela, M.: Cognitive Analysis Techniques in Business Planning and Decision Support Systems. In: Rutkowski, L., Tadeusiewicz, R., Zadeh, L.A., Żurada, J.M. (eds.) ICAISC 2006. LNCS (LNAI), vol. 4029, pp. 1027–1039. Springer, Heidelberg (2006)
136. Tadeusiewicz, R., Ogiela, L., Ogiela, M.R.: The automatic understanding approach to systems analysis and design. Elsevier, International Journal of Information Management 28, 38–48 (2008)
137. Tanaka, E.: Theoretical aspects of syntactic pattern recognition. Pattern Recognition 28, 1053–1061 (1995)
138. Tulving, E.: Precis of elements of episodic memory. The Behavioral and Brain Sciences 7, 223–268 (1984)
139. Tversky, A.: Elimination by aspects: A theory of choice. Psychological Review 79, 281–299 (1972)
140. von Neumann, J.: The Principles of Large-Scale Computing Machines. Annals of History of Computers 3(3), 263–273 (1946)
141. von Neumann, J.: Theory of Self-Reproducing Automata (edited and completed by A. Burks). University of Illinois Press, Urbana (1966)

142. Wang, Y.: On Cognitive Informatics, Brain and Mind: A Transdisciplinary. Journal of Neuroscience and Neurophilosophy 4(2), 151–167 (2003)

143. Wang, Y.: The Theoretical Framework of Cognitive Informatics. International Journal of Cognitive Informatics and Natural Intelligence 1(1), 1–27 (2007)

144. Wang, Y.: The Cognitive Processes of Formal Inferences. International Journal of Cognitive Informatics and Natural Intelligence 1(4), 75–86 (2007)

145. Wang, Y., Johnston, R., Smith, M. (eds.): Cognitive Informatics: Proceedings 1st IEEE International Conference (ICCI 2002). IEEE CS Press, Calgary (2002)

146. Wang, Y., Kinsner, W.: Recent Advances in Cognitive Informatics. IEEE Transactions on Systems, Man, and Cybernetics (Part C) 36(2), 121–123 (2006)

147. Wang, Y., Wang, Y., Patel, S., Patel, D.: A Layered Reference Model of the Brain (LRMB). IEEE Transactions on Systems, Man, and Cybernetics (Part C) 36(2), 124–133 (2006)

148. Wang, Y., Zhang, D., Latombe, J.C., Kinsner, W. (eds.): Proc. 7th IEEE International Conference on Cognitive Informatics (ICCI 2008). IEEE CS Press, Stanford University, CA, USA (2008)

149. Whewell, W.: History of Scientific Ideas. J.W.Parker & Son, London (1858)

150. Wiener, N.: Cybernetics. MIT Press, Cambridge (1948)

151. Wiener, N.: The Extrapolation. Interpolation and Smoothing of Stationary Time Series. Wiley, New York (1949)

152. Wilson, R.A., Keil, F.C.: The MIT Encyclopedia of the Cognitive Sciences. MIT Press (2001)

153. Yao, Y., Shi, Z., Wang, Y., Kinsner, W. (eds.): Proc. 5th IEEE International Conference on Cognitive Informatics (ICCI 2006). IEEE CS Press, Beijing (2006)

154. Zadeh, L.A.: Information and Control. Fuzzy Sets 8, 338–353 (1965)

155. Zadeh, L.A.: Fuzzy Sets and Systems. In: Fox, J. (ed.) Systems Theory, pp. 29–37. Polytechnic Press, Brooklyn (1965)

156. Zadeh, L.A.: Fuzzy logic, neural networks, and soft computing. Communications of the ACM 37(3), 77–84 (1994)

157. Zadeh, L.A.: Toward human level machine intelligence-Is it achievable? In: Proc. 7th International Conference on Cognitive Informatics (ICCI 2008). IEEE CS Press, Stanford University, CA (2008)

158. Zajonc, R.B.: On the Primacy of Affect. American Psychologist 39, 117–123 (1984)

159. Zhang, D., Wang, Y., Kinsner, W. (eds.): Proc. 6th IEEE International Conference on Cognitive Informatics (ICCI 2007). IEEE CS Press, Lake Tahoe (2007)

160. Zhong N., Raś Z.W., Tsumoto S., Suzuki E. (Eds.): Foundations of Intelligent Systems. In: 14th International Symposium, ISMIS 2003, Maebashi City, Japan (2003)

Internet Sources

161. http://eecs.vanderbilt.edu/cis/crl/index.shtml
162. http://www.alanturing.htm
163. http://www.AlfredNorthWhitehead.htm
164. http://www.AllenNewell.BiographicalMemoirs.htm
165. http://www.ArystotelesBiografia.htm
166. http://www.AugustDeMorgan.htm
167. http://www.BertrandRussell.htm
168. http://www.BoolePortraits.htm
169. http://www.cogniron.org
170. http://www.Computerworld–Ojcowieinformatyki.htm
171. http://www.Cyc.htm
172. http://www.DavidH_Hubel.htm
173. http://www.D_O_Hebb.htm
174. http://www/EDWARDTHORNDIKE.htm
175. http://www.EulerPortraits.htm
176. http://www.Fechner.htm
177. http://www.gatech.edu/
178. http://www.GottfriedWilhelmLeibniz.htm
179. http://www.HelmholtzPortraits.htm
180. http://www.HerbertSimon.htm
181. http://www.ImmanuelKant.htm
182. http://www.JohnMcCarthy.htm
183. http://www.LotfiZadeh.htm
184. http://www.LudwigWittgenstein.htm
185. http://www.lullianarts.net/cont.htm
186. http://www.MarvinMinsky.htm
187. http://www.MillJ_S.htm
188. http://www.NoamChomskyBiography.htm
189. http://www.nrl.navy.mil/aic/iss/aas/octavia.php
190. http://www.phys.uni.torun.pl/~duch
191. http://www.Porphyry.htm
192. http://www.RamonLlullus.htm
193. http://www.robonaut.jsc.nasa.gov/default.asp
194. http://www.TorstenN_Wiesel.htm
195. http://www.toyota.co.jp/en/special/robot/
196. http://www.Von_NeumannPortraits.htm
197. http://www.Wiener_NorbertPortraits.htm
198. http://www.WILHELMWUNDT.htm
199. http://www.William_Grey_Walter.htm

Index

A

Aristotle 1, 2, 4, 29
Ars Magna 2, 3, 4
Artificial intelligence 15, 27, 34

B

Behaviour 8, 12, 13, 14
Biometric analysis 86, 88, 91, 97
Biometric identification 85
Biometric systems 88
Brain 14, 19, 24

C

Characteristica Universalis 3
Cognition 5, 6, 7, 8, 9, 10, 15
Cognitive analysis 12, 32, 37, 49, 51, 91
Cognitive categorization 55
Cognitive computers 26
Cognitive decision systems 54
Cognitive informatics 19, 20, 29, 38, 107
Cognitive models
    Cognitive computer models 26
    Cognitive machine models 27
    The cognitive model of awareness
        and machine cognition 29
    The Cognitive Model of Memory
        (CMM) 27
    The Information-Matter-Energy
        (IME) 20
    The Information-Matter-Energy +
        Intelligence (IME+I) 22
    The Layered Reference Model of the
        Brain (LRMB) 23
    The model based on the structure of
        the natural mind 25
    The model of human perception
        processes 26

    The model of information
        representation in the
        human brain 24
    The natural intelligence model 26
    The Neural Informatics model (NeI)
        24
    The Object-Attribute-Relation
        (OAR) 24
Cognitive resonance 32, 35, 52
Cognitive robots 99, 103, 104, 106
Cognitive science 1, 6, 7, 14
Cognitive systems 99
Computational linguistics 4
Consciousness 8, 10
Cybernetics 6, 9, 19

D

Denotational mathematics 16
Differentiae 1

E

Emotions 10, 11
E-UBIAS systems 89, 97, 107

F

Formal grammar 36, 39
Formal language 32

G

Genus 1
Gottfried Wilhelm Leibniz 3

H

Hand bones 88
Hermann von Helmholtz 5

Human behaviour   8
Human intelligence   26

I

Image analysis systems   75
Information systems   32, 51
Isagoga, 1

J

John McCarthy   6
John Stuart Mill   4
John von Neumann   6

L

Limbic system   11
Lotfi A. Zadeh   6

M

Marvin Minsky   6
Mathematical linguistics   58, 76
Medical images   49
Mind   8

N

Noam Chomsky   6
Norbert Wiener   6

P

Plato   1, 2
Porphyry   1
Porphyry's diagram   2

R

Ramon Llull   2, 3, 4

S

Semantic analysis   36, 38, 67, 72, 96
Semantic categorization   35, 41, 43, 48
Semantic reasoning   59
Speech recognition   44
Spirit   2
Steven Abbott   3
Strategic analysis   61

U

UBIAS systems   75, 76, 78, 81,
                 82, 85
UBMSS systems   61, 66, 71
Understanding processes   7

Y

Yanis Dambergs   3

Cognitive Systems Monographs

Edited by R. Dillmann, Y. Nakamura, S. Schaal and D. Vernon

Vol. 1: Arena, P.; Patanè, L. (Eds.)
Spatial Temporal Patterns for
Action-Oriented Perception
in Roving Robots
425 p. 2009 [978-3-540-88463-7]

Vol. 2: Ivancevic, T.T.; Jovanovic, B.;
Djukic, S.; Djukic, M.; Markovic, S.
Complex Sports Biodynamics
326 p. 2009 [978-3-540-89970-9]

Vol. 3: Magnani, L.
Abductive Cognition
534 p. 2009 [978-3-642-03630-9]

Vol. 4: Azad, P.
Visual Perception for Manipulation
and Imitation in Humanoid Robots
270 p. 2009 [978-3-642-04228-7]

Vol. 5: de Aguiar, E.
Animation and Performance Capture
Using Digitized Models
168 p. 2010 [978-3-642-10315-5]

Vol. 6: Ritter, H.; Sagerer, G.;
Dillmann, R.; Buss, M.:
Human Centered Robot Systems
216 p. 2009 [978-3-642-10402-2]

Vol. 7: Levi, P.; Kernbach, S. (Eds.):
Symbiotic Multi-Robot Organisms
467 p. 2010 [978-3-642-11691-9]

Vol. 8: Christensen, H.I.;
Kruijff, G.-J.M.; Wyatt, J.L. (Eds.):
Cognitive Systems
491 p. 2010 [978-3-642-11693-3]

Vol. 9: Hamann, H.:
Space-Time Continuous Models of Swarm
Robotic Systems
147 p. 2010 [978-3-642-13376-3]

Vol. 10: Allerkamp, D.:
Tactile Perception of Textiles in a
Virtual-Reality System
120 p. 2010 [978-3-642-13973-4]

Vol. 11: Vernon, D.; von Hofsten, C.; Fadiga, L.:
A Roadmap for Cognitive Development
in Humanoid Robots
227 p. 2010 [978-3-642-16903-8]

Vol. 12: Ivancevic, T.T.; Jovanovic, B.;
Jovanovic, S.; Djukic, M.; Djukic, N.;
Lukman, A.:
Paradigm Shift for Future Tennis
375 p. 2011 [978-3-642-17094-2]

Vol. 13: Bardone, E.:
Seeking Chances
168 p. 2011 [978-3-642-19632-4]

Vol. 14: Jakimovski, B.:
Biologically Inspired Approaches for
Locomotion, Anomaly Detection and
Reconfiguration for Walking Robots
201 p. 2011 [978-3-642-22504-8]

Vol. 15: Der, R.; Martius, G.; Pfeifer, R.:
The Playful Machine
336 p. 2012 [978-3-642-20252-0]

Vol. 16: Zacharias, F.:
Knowledge Representations for Planning
Manipulation Tasks
143 p. 2012 [978-3-642-25181-8]

Vol. 17: Ogiela, L.; Ogiela, M.R.:
Advances in Cognitive Information Systems
120 p. 2012 [978-3-642-25245-7]

Printed by Publishers' Graphics LLC
MO20120828-106